PHYLOGENETIC SYSTEMATICS OF IGUANINE LIZARDS
A COMPARATIVE OSTEOLOGICAL STUDY

University of California Publications

ZOOLOGY
Volume 118

Phylogenetic Systematics of Iguanine Lizards

A Comparative Osteological Study

by Kevin de Queiroz

University of California Press

591.08
C153u
v.118
1981

Phylogenetic Systematics of Iguanine Lizards

A Comparative Osteological Study

by Kevin de Queiroz

A Contribution from the Museum of Vertebrate Zoology
of the University of California at Berkeley

UNIVERSITY OF CALIFORNIA PRESS
Berkeley • Los Angeles • London

UNIVERSITY OF CALIFORNIA PUBLICATIONS IN ZOOLOGY

Editorial Board: Peter B. Moyle, James L. Patton,
Donald C. Potts, David S. Woodruff

Volume 118
Issue Date: December 1987

UNIVERSITY OF CALIFORNIA PRESS
BERKELEY AND LOS ANGELES, CALIFORNIA

UNIVERSITY OF CALIFORNIA PRESS, LTD.
LONDON, ENGLAND

ISBN 0-520-09730-0
LIBRARY OF CONGRESS CATALOG CARD NUMBER: 87-24594

© 1987 BY THE REGENTS OF THE UNIVERSITY OF CALIFORNIA
PRINTED IN THE UNITED STATES OF AMERICA

Library of Congress Cataloging-in-Publication Data
De Queiroz, Kevin.
    Phylogenetic systematics of iguanine lizards: a comparative
osteological study / by Kevin de Queiroz.
        p.   cm. — (University of California publications in zoology:
v. 118)
    Bibliography: p.
    ISBN 0-520-09730-0 (alk. paper)
    1. Iguanidae—Classification. 2. Iguanidae—Evolution.
3. Iguanidae—Anatomy. 4. Anatomy, Comparative. 5. Reptiles—
Classification. 6. Reptiles—Evolution. 7. Reptiles—Anatomy.
I. Title. II. Series.
QL666.L25D4 1987
597.95—dc 19   87-24594
                                                                        CIP

Contents

List of Illustrations, vii
List of Tables, x
Acknowledgments, xi
Abstract, xii

INTRODUCTION   1
   Historical Review, 1
   Goals of This Study, 10

MATERIALS AND METHODS   13
   Specimens, 13
   Phylogenetic Analysis, 13
   Basic Taxa, 14
   The Problem of Variation, 14
   Construction of Branching Diagrams, 16

IGUANINE MONOPHYLY   18

COMPARATIVE SKELETAL MORPHOLOGY   21
   Skull Roof, 21
   Palate, 39
   Braincase, 44
   Mandible, 49
   Miscellaneous Head Skeleton, 59
   Axial Skeleton, 69
   Pectoral Girdle and Sternal Elements, 81
   Pelvic Gridle, 86
   Limbs, 89
   Osteoderms, 89

NONSKELETAL MORPHOLOGY   92
   Arterial Circulation, 92
   Colic Anatomy, 93
   External Morphology, 94

SYSTEMATIC CHARACTERS 100
   Skeletal Characters, 100
   Nonskeletal Characters, 104

CHARACTER POLARITIES AND THE PHYLOGENETIC INFORMATION
   CONTENT OF CHARACTERS 106

ANALYSIS OF PHYLOGENETIC RELATIONSHIPS 117
   Preliminary Analysis, 117
   Lower Level Analysis, 122

PHYLOGENETIC CONCLUSIONS 130
   Preferred Hypothesis of Relationships, 130
   Character Evolution within Iguaninae, 130

COMPARISONS WITH PREVIOUS HYPOTHESES 132

DIAGNOSES OF MONOPHYLETIC GROUPS OF IGUANINES . . . 135
   Iguaninae Bell 1825, 135
     Dipsosaurus Hallowell 1854, 141
     Brachylophus Wagler 1830, 143
     Iguanini Bell 1825, 145
       Ctenosaura Wiegmann 1828, 146
       Sauromalus Duméril 1856, 157
       Amblyrhynchina, new taxon, 160
         Amblyrhynchus Bell 1825, 163
         Conolophus Fitzinger 1843, 165
       Iguanina Bell 1825, 167
         Iguana Laurenti 1768, 168
         Cyclura Harlan 1824, 170

Appendix I: Specimens Examined, 175
Appendix II: Polarity Determination Under Uncertain Outgroup Relationships, 179
Appendix III: Polarity Determination for Lower Level Analysis, 185
Appendix IV: Polarity Reevaluation for Lower Level Analysis, 187
Literature Cited, 191

List of Illustrations

FIGURES

1. "The phylogeny and relationships of North American iguanid genera," after Mittleman (1942), 6
2. "Grouping and possible phylogeny of the genera of iguanids occurring in the United States," after H. M. Smith (1946), 7
3. "Phylogenetic relationships of the Madagascar Iguanidae and the genera of iguanine lizards," after Avery and Tanner (1971), 9
4. Etheridge's phylogeny of the Iguanidae, 11
5. Skull of *Brachylophus vitiensis*, 22
6. Skull and mandible of *Brachylophus vitiensis*, 23
7. Posteroventral views of iguanine premaxillae, 24
8. Dorsal views of the preorbital portions of iguanine skulls, 25
9. Dorsal views of the skulls of *Cyclura cornuta* and *Sauromalus obesus*, 27
10. Posterodorsal views of the anterior orbital regions of *Brachylophus fasciatus* and *Conolophus pallidus*, 28
11. Dorsal view of the skull of *Amblyrhynchus cristatus*, 29
12. Ventral views of iguanine frontals, 31
13. Dorsal views of the parietals in an ontogenetic series of *Iguana iguana*, 34
14. Lateral view of the skull of *Ctenosaura similis*, 36
15. Lateral views of the posterolateral corners of iguanine skulls, 38
16. Posterodorsal views of disarticulated right palatines of *Iguana delicatissima* and *Conolophus subcristatus*, 40
17. Posterodorsal views of the right orbits of five iguanines and *Morunasaurus annularis*, 41
18. Ventral view of the skull of *Iguana delicatissima*, 43
19. Anterolateral views of the left orbitosphenoids in an ontogenetic series of *Iguana iguana*, 45
20. Ventral views of the posterior portion of the palate and anterior portion of the braincase of *Sauromalus varius* and *Amblyrhynchus cristatus*, 46
21. Ventral views of iguanine neurocrania, 47
22. Lateral views of the right mandibles of *Iguana delicatissima* and *Amblyrhynchus cristatus*, 50
23. Lingual views of the left mandibles of three iguanines, 51
24. Lateral views of the right mandibles of *Conolophus pallidus* and *Cyclura cornuta*, 52

List of Illustrations

25. Lateral views of the right mandibles of *Iguana delicatissima, Sauromalus obesus*, and *Amblyrhynchus cristatus*, 53
26. Lateral views of the right mandibles of *Dipsosaurus dorsalis, Brachylophus vitiensis*, and *Iguana iguana*, 55
27. Medial views of the left mandibles of *Iguana delicatissima* and *Conolophus subcristatus*, 56
28. Dorsal views of the posterior ends of the right mandibles in ontogenetic series of *Ctenosaura hemilopha* and *Amblyrhynchus cristatus*, 57
29. Dorsal views of the posterior ends of the right mandibles in an ontogenetic series of *Dipsosaurus dorsalis*, 58
30. Lingual views of left maxillary teeth of four iguanines and *Basiliscus plumifrons*, 62
31. Hypothetical character phylogeny for the iguanine pterygoid tooth patch, 65
32. Corneal view of the left scleral ring of *Ctenosaura similis*, 67
33. Ventral views of the iguanine hyoid apparati, 68
34. Twentieth presacral vertebra of *Brachylophus vitiensis*, 70
35. Lateral views of the twentieth presacral vertebrae of *Sauromalus obesus* and *Ctenosaura pectinata*, 72
36. Dorsolateral views of the twentieth presacral vertebrae of *Dipsosaurus dorsalis* and *Sauromalus obesus*, 73
37. Dorsal views of caudal vertebrae of *Dipsosaurus dorsalis* from different regions of the tail, 76
38. Lateral views of the ninth caudal vertebrae of *Dipsosaurus dorsalis* and *Iguana iguana*, 79
39. Presacral and sacral vertebrae and ribs of *Dipsosaurus dorsalis* in ventral view, 80
40. Pectoral girdles of three iguanines, 82
41. Dorsal views of the pelvic girdles of *Sauromalus obesus* and *Ctenosaura pectinata*, 86
42. Bones of the anterior limb of *Brachylophus fasciatus*, 87
43. Right hind limb skeleton of *Brachylophus fasciatus*, 88
44. Right tarsal region of *Brachylophus fasciatus*, 90
45. Anterodorsal views of pedal digit II of three iguanines, 97
46. Minimum-step cladograms for eight basic taxa of iguanines resulting from a preliminary analysis of 29 characters, 119
47. Alternative interpretations of character transformation for homoplastic characters on a minimum-step cladogram, 121
48. Alternative interpretations of character transformation for homoplastic characters on a minimum-step cladogram, 122
49. Minimum-step cladograms resulting from an analysis of 26 characters in a subset of iguanines, 127
50. Consensus cladogram for the three cladograms illustrated in Figure 49, 128
51. Phylogenetic relationships within Iguaninae, according to the present study, 131
52. Geographic distribution of *Dipsosaurus*, 141
53. Geographic distribution of *Brachylophus*, 144

54. Geographic distribution of *Ctenosaura*, 147
55. Cladogram illustrating phylogenetic relationships within *Ctenosaura*, 154
56. Geographic distribution of *Sauromalus*, 158
57. Geographic distribution of Amblyrhynchina (*Amblyrhynchus* and *Conolophus*), 161
58. Geographic distribution of *Iguana*, 169
59. Geographic distribution of *Cyclura*, 171
60. All nine possible fully resolved cladogram topologies for four unspecified outgroups and an ingroup, 179
61. Dendrograms corresponding with the nine cladograms in Figure 60 after each is rerooted at the outgroup node, 180
62. Examples of polarity inferences for different arrangements of outgroup character state distributions, 182
63. All possible cladogram topologies for two unspecified outgroups and an ingroup before and after rerooting at the outgroup node, 185
64. All possible cladogram topologies for two unspecified near outgroups, one more remote outgroup, and an ingroup before and after rerooting at the outgroup node, 186

PLATE

1. Lateral and dorsal views of the skull of *Amblyrhynchus cristatus*, 91

List of Tables

1. The iguanine genera, 2
2. Position of the parietal foramen, 32
3. Numbers of premaxillary teeth, 60
4. Numbers of presacral vertebrae, 71
5. Distributions of character states of 95 characters among four outgroups to iguanines and the polarities that can be inferred from them, 108
6. Distributions of character states of 95 characters among eight iguanine taxa, 112
7. Distributions of character states of 29 characters used in the preliminary analysis, 118
8. Polarity inferences for lower-level analysis, using *Brachylophus* and *Dipsosaurus* as outgroups, 124
9. Distributions of character states of 26 characters among six taxa within Iguanini, 125
10. Distributions of character states of 19 characters among basic taxa within *Ctenosaura* (in the broad sense) and three close and two more distant outgroups, 153
11. Summary of polarity inferences for seven cases of character-state distribution among four outgroups of uncertain relationships to the ingroup, 181
12. Summary of polarity inferences for four cases of character-state distribution among two outgroups of uncertain relationships to the ingroup, 185
13. Summary of polarity inferences for six cases of character-state distribution among two near outgroups whose precise relationships to the ingroup are unresolved, and one more remote outgroup exhibiting a fixed character state, 187

Acknowledgments

Many people have helped me toward the completion of this study in ways big and small. Over the years I have undoubtedly forgotten the contributions of some of them, and I apologize for this. Of those I have not forgotten, I want to thank the following people for lending me specimens under their care: Pere Alberch, Walter Auffenberg, James Berrian, Robert Bezy, Steven Busack, Joseph Collins, Ronald Crombie, Mark Dodero, Robert Drewes, William Duellman, Anne Fetzer, George Foley, Harry Greene, L. Lee Grismer, W. Ronald Heyer, J. Howard Hutchinson, Charles Meyers, Peter Meylan, Mark Norell, Gregory Pregill, Jose Rosado, Albert Schwartz, Jens Vindum, Van Wallach, John Wright, George Zug, Richard Zweifel, and especially Jay Savage and Richard Etheridge whose collections provided the majority of the specimens examined in this study.

I am also grateful to various teachers, friends, and colleagues who helped my ideas on systematics and iguanine biology unfold through countless discussions: Troy Baird, Aaron Bauer, Theodore Cohn, Michael Donoghue, Richard Estes, Richard Etheridge, Jacques Gauthier, Eric Gold, David Good, George Gorman, Scott Lacour, Eric Lichtwardt, James Melli, Sheldon Newberger, Mark Norell, Michael Novacek, David Wake, and André Wyss. Linda Condon-Howe, Charles Crumly, Sanae and John Moorehead, Douglas Preston, Doris Taylor, and the late Kenneth Miyata generously provided lodging while I was visiting museums. Richard Estes, Richard Etheridge, Darrel Frost, Gregory Pregill, David Wake, and Edward Warren, provided valuable comments on earlier versions of the manuscript. David Cannatella and Rose Anne White greatly assisted in the preparation of camera-ready-copy.

Finally, I want to give special thanks to Karen Sitton for providing emotional support in her unique and charming way and to Richard Etheridge and Richard Estes for their influence on both my academic and personal development.

This study partially fulfilled the requirements of a Master's degree in Zoology at San Diego State University, but was completed at the University of California, Berkeley. The research and preparation of the manuscript were supported in part by a grants from the Society of Sigma Xi, the San Diego State University Department of Zoology, the Theodore Roosevelt Memorial Fund of the American Museum of Natural History, and the Graduate Student Research Allocation Fund of the Department of Zoology, University of California at Berkeley.

Abstract

Iguaninae is a monophyletic taxon of tetrapodous squamates (lizards) that can be distinguished from other iguanians by at least five synapomorphies. Skeletal variation within Iguaninae is described and forms the basis of systematic characters used to determine phylogenetic relationships among eight basic taxa, the currently recognized iguanine genera. Evolutionary character polarities are determined by comparison with four closely related taxa, basiliscines, crotaphytines, morunasaurs, and oplurines.

The distributions of derived characters among iguanine taxa suggest that: (1) Either *Brachylophus* or *Dipsosaurus* is the sister group of the remaining iguanines (Iguanini). (2) *Dipsosaurus* is a monophyletic taxon diagnosed by at least six synapomorphies. (3) *Brachylophus* is a monophyletic taxon diagnosed by at least eight synapomorphies. (4) Iguanini, containing *Amblyrhynchus, Conolophus, Ctenosaura, Cyclura, Iguana,* and *Sauromalus,* is a new monophyletic taxon diagnosed by at least three synapomorphies. (5) Within Iguanini, the relationships among four taxa-*Ctenosaura, Sauromalus,* Amblyrhynchina, and Iguanina-are unresolved. (6) *Ctenosaura* is a monophyletic taxon diagnosed by at least three synapomorphies. (7) *Enyaliosaurus* is monophyletic, but it is a subgroup of *Ctenosaura* rather than a separate taxon. If *Enyaliosaurus* is separated from *Ctenosaura,* then *Ctenosaura* is not monophyletic. (8) *Sauromalus* is a monophyletic taxon diagnosed by at least 24 synapomorphies, many of which are convergent in *Amblyrhynchus.* (9) Amblyrhynchina is a new monophyletic taxon containing the Galápagos iguanas *Amblyrhynchus* and *Conolophus,* and is diagnosed by at least 11 synapomorphies. (10) *Amblyrhynchus* is a monophyletic taxon diagnosed by at least 28 synapomorphies and is perhaps the most divergent iguanine from the most recent common ancestor of all of them. Many of the unique features of *Amblyrhynchus* appear to be related to its unique natural history. (11) *Conolophus* is a monophyletic taxon diagnosed by at least eight synapomorphies and cannot, therefore, be considered ancestral to *Amblyrhynchus.* (12) Iguanina is a new monophyletic taxon composed of *Iguana* and *Cyclura* and is diagnosed by at least three synapomorphies. (13) *Iguana* is a monophyletic taxon diagnosed by at least seven synapomorphies. (14) Monophyly of *Cyclura* is a problem in need of further study. Although three ostensible synapomorphies support monophyly of *Cyclura,* other derived characters suggest that some *Cyclura* shared a more recent common ancestor with *Iguana* than with other *Cyclura.*

Summaries of Iguaninae and its monophyletic subgroups down to the level of the eight basic taxa are provided; each summary includes the type of the taxon, etymology of the taxon name, a phylogenetic definition, geographic distribution, a list of diagnostic synapomorphies, the fossil record, and various comments.

INTRODUCTION

Containing approximately 55 genera and more than 600 species, Iguanidae is one of the largest families of lizards. Its members occur primarily in the New World, from southern Canada to austral South America including the Galápagos Archipelago and much of the West Indies. Iguanids also occur on the island of Madagascar and in the Comores Archipelago in the western Indian Ocean, and on the Fiji and Tonga island groups in the southwestern Pacific.

For over 100 years, systematists have attempted to discover the pattern of interrelationships among the genera in the family Iguanidae, but, because of the bewildering morphological diversity within this family, the task is far from complete. Nevertheless, many systematists have recognized suprageneric groups of iguanids (e.g., Wagler, 1830; Duméril and Bibron, 1837; Fitzinger, 1843; Gray, 1845; Cope, 1886, 1900; Boulenger, 1890; H. M. Smith, 1946; Savage, 1958; Etheridge, 1959, 1964a). One of the earliest of these suprageneric groups to be recognized consists of the genera currently known informally as iguanines. This assemblage is also one of the most readily diagnosed on the basis of uniquely derived features. As currently conceived, there are eight genera and 31 species of iguanines (Etheridge, 1982). The iguanine genera are listed in Table 1, which also gives the number of included species, their habits, and the geographic distribution for each genus.

HISTORICAL REVIEW

The concept of an iguanine group is remarkably old, predating the publication of Darwin's *Origin of Species* (1859). This accomplishment is even more surprising when one realizes that all iguanines are native to regions far from western Europe, where systematists were developing the concept of an iguanine group. These systematists undoubtedly had few specimens at hand, and must have relied heavily on each others' character descriptions. Although I have been unable to see all of the potentially relevant literature, I attempt to trace and summarize the history of iguanine higher systematics.

The Eighteenth Century. Although the eighteenth century was an important one for biological systematics as a whole, it was not so important for iguanine systematics. A convenient date to begin a historical discussion of iguanine systematics is 1758, when Linnaeus published the tenth edition of his *Systema Naturae*, the starting point of zoological nomenclature. Linnaeus himself was neither interested in nor fond of the "lower" tetrapods. He placed all tetrapodous squamates in two genera, one of which

TABLE 1. The Iguanine Genera

Genus (common name)	Number of Species	Habits	Geographic Distribution
Amblyrhynchus Bell 1825 (Marine Iguanas)	1	Terrestrial, saxicolous, semimarine	Rocky coasts of various islands of the Galápagos Archipelago, Ecuador.
Brachylophus Wagler 1830 (Banded Iguanas)	2	Arboreal	Various South Pacific islands of the Fiji and Tonga groups.
Conolophus Fitzinger 1843 (Galápagos Land Iguanas)	2	Terrestrial	Islands of the Galápagos Archipelago, Ecuador.
Ctenosaura Wiegmann 1828 (Spiny-tailed Iguanas)	9	Terrestrial, arboreal	Lowlands of México and Central America, including various offshore islands, as far south as Panamá.
Cyclura Harlan 1824 (West Indian Ground Iguanas)	8	Terrestrial	The Bahama Islands; Cayman Islands; Navassa, Mona, and Anegada islands; and Cuba, Hispaniola, and Jamaica, and their nearby islets.
Dipsosaurus Hallowell 1854 (Desert Iguanas)	1	Terrestrial	Deserts of the southwestern United States, northwestern mainland México, Baja California, and various islands in the Gulf of California.
Iguana Laurenti 1768 (Green Iguanas)	2	Arboreal	Lowlands of México, Central America, and South America to southern Brazil and Paraguay; in the Caribbean northward through the Lesser Antilles to the Virgin Islands.
Sauromalus Duméril 1856 (Chuckwallas)	6	Terrestrial, saxicolous	Deserts of the southwestern United States, northwestern mainland México, Baja California, and various islands in the Gulf of California.

Sources: Etheridge (1982) and Gibbons (1981).

contained *Lacerta iguana* (=*Iguana iguana*), the single known iguanine, and animals now placed in at least 12 different families, including crocodilians and amphibians. He considered them to be "foul and loathsome animals" (Linnaeus, 1758, translated in Goin et al., 1978). At the close of the eighteenth century only three of the currently recognized iguanine species (now placed in two genera) had been described, giving the systematists of that century, such as Laurenti (1768) and Lacépède (1788), little of a group to recognize.

The Nineteenth Century. Major advances in iguanine systematics came during the nineteenth century. Many important natural histories and systems or classifications of squamates appeared during these years, and by 1856 all of the currently recognized iguanine genera had been described.

The concept of a natural iguanine taxon emerged during the first half of the nineteenth century. Most of the authors of classifications published during this period recognized a close relationship among at least some of the iguanine genera. Those that did not recognize a complete and exclusive group for the iguanines known at the time failed to do so for one or both of two reasons. Brongniart (1805), Latreille (1825), Fitzinger (1826, 1843), Wagler (1830), and Duméril and Bibron (1837) grouped all the known iguanines together, but included some noniguanines with them. Although all the iguanines were sometimes placed together as part of a continuous list, it is not evident that they were considered to form their own subgroup within some larger group. Other authors such as Daudin (1805), Merrem (1820), Cuvier (1829, 1831), and Wagler (1830) failed to place all iguanines in a single group. Daudin, Cuvier, and Wagler included *Brachylophus* with the agamids, while Merrem did the same for *Ctenosaura*.

At least three authors can truly be said to have recognized an iguanine group before 1850. I have two criteria for determining the true recognition of an iguanine group. First, all of the iguanine taxa known to the author (or at least all those listed in the classification) were included in the group; and second, no other taxa were included. Cuvier's (1817) "*Les Iguanes proprement dits*" consisted of what are now *Iguana iguana*, *I. delicatissima*, *Cyclura cornuta*, and *Brachylophus fasciatus*, although he later removed *Brachylophus* and placed it among the agamids (Cuvier, 1829, 1831). Wiegmann (1834) placed only the genera *Cyclura*, *Ctenosaura*, *Iguana*, *Brachylophus*, and *Amblyrhynchus* in his family Dendrobatae, Tribus II, b, ***, ß. Like many of his contemporaries, Wiegmann constructed his classification as a hierarchy of sets and subsets that would also function as a key.

The most fully developed early concept of an iguanine group appears to have been that of Gray (1831a, 1845). In 1831, Gray placed all known iguanines (equivalent to what are now 10 species in five genera) by themselves in a single genus, *Iguana*. Fourteen years later, he recognized nine different iguanine genera. Because these nine genera (again equivalent to five modern genera) formed one entire set in his hierarchical classification, it is evident that Gray still recognized the unity of the iguanine group.

Progress in iguanine systematics, though less rapid than in the previous fifty years, continued through the second half of the nineteenth century. The last two iguanine genera that are still recognized, *Dipsosaurus* and *Sauromalus*, were described, but at first they were not explicitly included with the rest of the iguanines in an exclusive group. The concept of an iguanine group, exclusive of *Dipsosaurus* and *Sauromalus*, was refined with more detailed anatomical descriptions. Beginning with Boulenger's (1885) monumental *Catalogue of the Lizards in the British Museum*, I undertake here a more detailed chronological treatment of the history of iguanine higher systematics.

Boulenger (1885) listed all of the genera that are now called iguanines in a nearly continuous sequence in his catalogue, reflecting their position in his key as those iguanids having femoral pores and the fourth toe longer than the third but lacking spines on the head and an enlarged occipital scale. Nevertheless, the distantly related *Hoplocercus* (Etheridge in Paull et al., 1976) breaks the continuity of the iguanines in the list, and, in terms of Boulenger's characters, some iguanines are closer to certain non-iguanine iguanids than to other iguanines. Boulenger did not explicitly delimit subgroups within Iguanidae or any other family, and we can only guess about his precise ideas concerning such relationships.

Cope (1886) appears to have been the first to use the name Iguaninae as a formal taxon for iguanine lizards. He further provided characters, both external and skeletal, by which members of this group could be distinguished from other iguanids. Cope's Iguaninae included *Cyclura*, *Ctenosaura*, *Cachryx*, *Brachylophus*, *Iguana*, *Conolophus*, and *Amblyrhynchus*, but failed to include *Dipsosaurus* and *Sauromalus*. The genera *Aloponotus* and *Metopoceros* were synonymized with *Cyclura*.

In response to Cope, Boulenger (1890) provided what he considered to be osteological evidence for the separation of *Metopoceros* and *Cyclura*, and briefly described the skulls of "the iguanoid lizards allied to *Iguana*." Except for the recognition of *Metopoceros* and the omission of *Cachryx*, the genera included in this discussion were the same as Cope's (1886) Iguaninae. *Dipsosaurus* and *Sauromalus* were again left out of the group.

Cope later (1900) greatly expanded his Iguaninae, and named two additional iguanid subfamilies, Anolinae and Basiliscinae. This new Iguaninae was a catch-all group for those iguanids that lacked midventrally continuous postxiphisternal inscriptional ribs, had simple clavicles, and lacked a left hepatopulmonary mesentery--in other words, those iguanids that lacked the distinctive features of anolines and basiliscines. Although this new Iguaninae was almost certainly an unnatural group, Cope recognized a slightly expanded version of his earlier (1886) Iguaninae as a discrete subset of his new and more inclusive group of the same name. This unnamed subset was characterized by the presence of femoral pores and of vertebrae with zygosphenal articulations. It contained *Dipsosaurus* and *Sauromalus* along with the genera included in his earlier Iguaninae; and it is therefore identical in generic content to the iguanine group as currently conceived.

The Twentieth Century. During the first three-fourths of the twentieth century, the concept of an iguanine group underwent considerable change. The efforts of nineteenth-century authors such as Cope and Boulenger seem to have been largely ignored, and at least two authors envisioned the ancestry of most other North American iguanids within iguanines. This idea seems to have resulted from the misconception that iguanines were "primitive" iguanids and were, therefore, potential ancestors of other iguanid taxa; the integrity of the group was deemphasized or completely overlooked. Nevertheless, by the mid-1960's the iguanines had been resurrected as a natural group, the same group that Cope (1900) had recognized at the turn of the century.

In his landmark paper on squamate systematics, *Classification of the Lizards*, Camp (1923) dealt primarily with the interrelationships of the lizard families. Nevertheless, his treatise contains scattered but intriguing comments on relationships at lower taxonomic levels. About the throat musculature of iguanines, he said:

> In the "Cyclura group" comprising the genera *Iguana, Amblyrhynchus, Ctenosaura, Brachylophus, Sauromalus,* and *Cyclura,* the superficial bundle [of the *M. mylohyoideus anterior*] is very specialized and consists of definitely directed fibers not connected with the skin. Detailed resemblances are present in this group which I have outlined in manuscript and which will not be repeated here. Suffice it to say that the group appears to be a natural one, on the basis of the musculature with close resemblances prevalent between *Sauromalus* and *Cyclura,* and *Ctenosaura* and *Brachylophus.* (Camp, 1923:371)

Unfortunately, the whereabouts of the manuscript mentioned in this passage are unknown to me.

Mittleman (1942) reviewed the genus *Urosaurus* and commented briefly on the relationships among the genera of North American iguanids, except *Anolis*. He implied that the North American iguanids formed a monophyletic group descended from *Ctenosaura* (Fig. 1) and that the similarities among *Ctenosaura, Dipsosaurus,* and *Sauromalus* were retained primitive features:

> *Dipsosaurus* is probably the most primitive of the North American *Iguanidae* (excepting *Ctenosaura*, which is properly a Central and South American form), and possesses several points in common with *Ctenosaura*, most easily observed of which is the dorsal crest; the genera further show their relationship in the similarity of the cephalic scutellation which is essentially simple, and shows no particular degree of differentiation. *Sauromalus* is considered a specialized offshoot of the *Crotaphytus*, or more properly, pre-*Crotaphytus* stock, by reason of its solid sternum, as well as the five-lobed teeth; the simple type of cephalic scalation indicates its affinity with the more primitive *Dipsosaurus-Ctenosaura* stock. (Mittleman, 1942:112-113)

H. M. Smith (1946:92) seemed to adopt a modified version of Mittleman's views on the phylogeny of North American iguanids (Fig. 2). His herbivore section (*Ctenosaura, Dipsosaurus,* and *Sauromalus*) was considered to be ancestral to the other North American Iguanidae, save *Anolis*, with *Sauromalus* hypothesized to share a more recent common ancestry with these other iguanids than with either *Ctenosaura* or *Dipsosaurus*. Smith's subsequent comments (1946:101), however, indicate that he recognized affinities of *Ctenosaura, Dipsosaurus,* and *Sauromalus* to iguanids occurring outside of the United

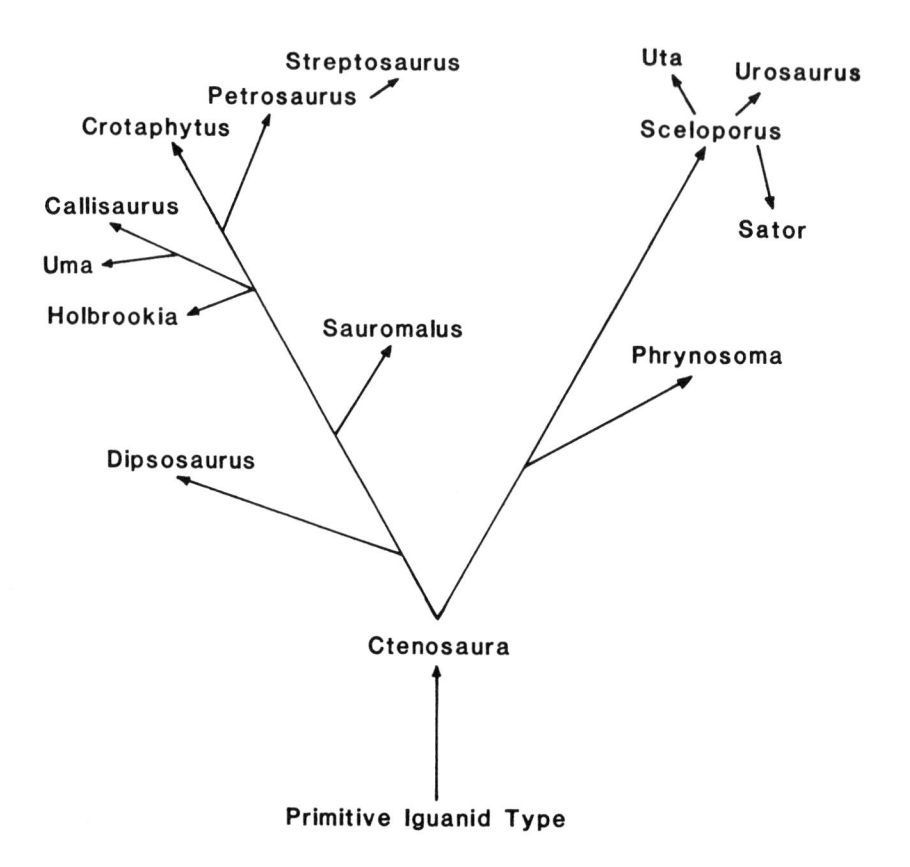

FIG. 1. "The phylogeny and relationships of North American iguanid genera," after Mittleman (1942:113).

States. In addition to the three genera found in or near the United States, Smith's herbivore section contained other "large, primitive iguanids," namely *Amblyrhynchus*, *Conolophus*, *Cyclura*, and *Iguana*. Smith's *Handbook* dealt with the lizards of the United States and Canada; those iguanines whose ranges did not enter this area were apparently omitted from his phylogram for convenience. In any case, Smith could not have considered his herbivore section to be monophyletic in the more restricted modern sense, since the group was considered to be ancestral to other North American iguanids.

Savage (1958) explicitly challenged Mittleman's (1942) implication that the North American iguanids formed a natural group:

> Insofar as can be determined at this time, the so-called Nearctic iguanids form two diverse groups that can only be distantly related. These two sections are

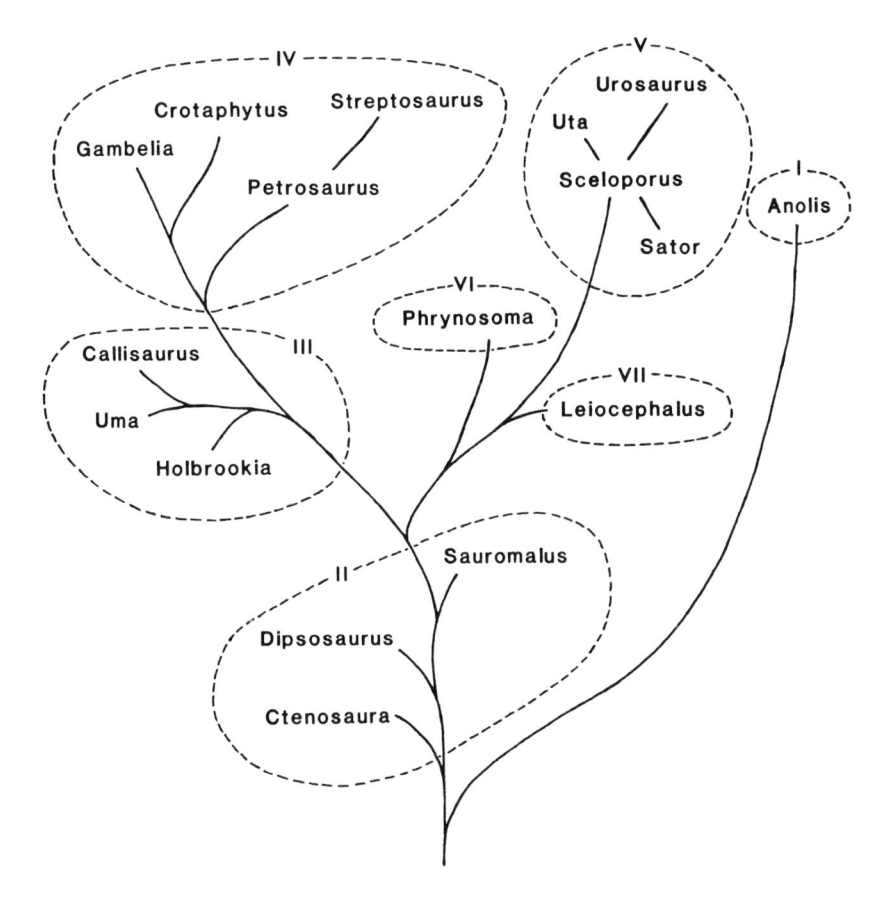

FIG. 2. "Grouping and possible phylogeny of the genera of iguanids occurring in the United States," after H. M. Smith (1946:92). Roman numerals apparently refer to the following: (I) leaf-toed section, (II) herbivore section, (III) sand-lizard section, (IV) rock-lizard section, (V) pored utiform section, (V) horned-lizard section, and (VII) poreless utiform section.

distinguished by marked differences in vertebral and nasal structures and include several genera not usually recognized as being allied to Nearctic forms. (Savage, 1958:48)

Savage's "iguanine line" contained *Amblyrhynchus*, *Brachylophus*, *Conolophus*, *Crotaphytus*, *Ctenosaura*, *Cyclura*, *Dipsosaurus*, *Enyaliosaurus* (=*Ctenosaura*, part), *Iguana*, and *Sauromalus*. This group was distinguished from the "sceloporine line" by two primary characters: the presence of accessory vertebral articulations, the zygosphenes and zygantra, and the possession of a relatively simple, S-shaped nasal passage with a concha present (*Dipsosaurus*-type of Stebbins, 1948). Other osteological and integumentary features characteristic of the majority of the genera in each line were also given.

The currently recognized iguanine group is based on the work of Etheridge. In his paper on the systematic relationships of sceloporine lizards, Etheridge (1964a) showed that the two primary characters used by Savage (1958) to diagnose the iguanines were actually more widespread within the Iguanidae, and were thus insufficient to diagnose the group. He listed four fundamental differences between *Crotaphytus* and Savage's other iguanines, and asserted that if *Crotaphytus* was considered to be an iguanine, no character or combination of characters could be used to diagnose that group. Once he removed *Crotaphytus* from the group, the iguanines were readily diagnosed by their unique caudal vertebrae. Except for his recognition of *Enyaliosaurus* as a genus separate from *Ctenosaura*, Etheridge's (1964a) concept of the iguanines is identical to that held today (Etheridge, 1982).

Despite the long history of iguanines as a recognized group and the great interest in many aspects of iguanine biology (e.g., Burghardt and Rand, 1982; Troyer, 1983), the interrelationships among the iguanine genera and the relationships of iguanines to other iguanians remain largely unknown. Commonly held beliefs are that *Ctenosaura* and *Cyclura* are closely related (Barbour and Noble, 1916; Bailey, 1928; Schwartz and Carey, 1977), and that the same is true of the Galápagos iguanas *Amblyrhynchus* and *Conolophus* (Heller, 1903; Eibl-Eibesfeldt, 1961; Thornton, 1971; Higgins, 1978). As mentioned above, Mittleman (1942) and H. M. Smith (1946) have offered dendrograms depicting their views on the relationships of the North American iguanines.

Recent studies have examined diverse data for clues about the interrelationships among the iguanine genera, but have met with limited success. Zug (1971) studied the arterial system of iguanids. He published shortest-connection networks for more than 40 iguanid genera, some based on his arterial characters and others based on characters obtained from the literature, most of which were osteological. Other shortest-connection networks constructed from data on arterial variation within various suprageneric assemblages of iguanids, including iguanines, were also presented. Nevertheless, Zug doubted the usefulness of his arterial characters in iguanid systematics, stating: "The arterial characters employed herein appear to be of minimal value in iguanid classification. At the intrafamilial level, they are disruptive and form groups of questionable zoogeographic unity" (Zug, 1971:21).

There has been but a single study in which the relationships among all known iguanine genera were sought, that of Avery and Tanner (1971). These authors provided descriptions of the iguanine skeleton, head and neck musculature, tongue, and hemipenes, and gave a number of osteological measurements. They based their hypothesis of relationships on mean length-width ratios of bones, assuming that "a difference of forty or less points between means of the same bone indicates a close relationship" (Avery and Tanner, 1971:67). Large numbers of such similarities were taken to indicate close phylogenetic relationship among taxa and were used in some unspecified way to construct a phylogenetic diagram (Fig. 3). Avery and Tanner examined small series (never more than five individuals of a single species), giving no consideration to allometric changes in the ratios that they used. I suspect that many of these ratios are correlated with a single

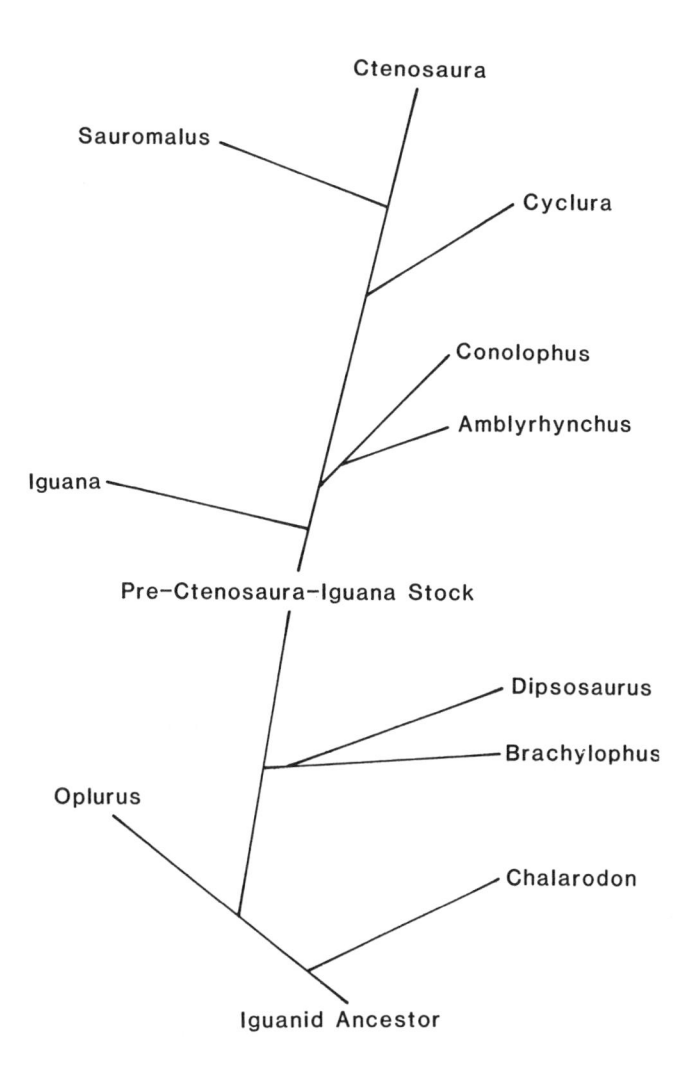

FIG. 3. "Phylogenetic relationships of the Madagascar Iguanidae and the genera of iguanine lizards," after Avery and Tanner (1971:71).

variable, size, and should not therefore be used as independent evidence for relationship. Furthermore, these authors made no attempt to assess the evolutionary polarity of their characters by comparison with other iguanids.

Karyological data on iguanines have been practically useless for systematic purposes. At the crude level of karyotypic analysis commonly applied to lizards, in which only numbers and sizes of chromosomes and their centromeric positions are determined, iguanines are conservative. All species of *Conolophus*, *Cyclura*, *Ctenosaura*, *Dipsosaurus*, and *Sauromalus* that have been studied possess a karyotype known to be widespread within Iguanidae and found in several other lizard families as well (Paull et al., 1976). Only *Iguana iguana* has been reported to differ from this seemingly primitive condition in that this species supposedly lacks one pair of microchromosomes (Cohen et al., 1967), but even this finding was contradicted in another study (Gorman et al., 1967; Gorman, 1973).

Iguanine relationships have only been studied superficially with relatively new and increasingly popular biochemical techniques. Gorman et al. (1971) presented evidence for close relationship among iguanines based on immunological studies of lactic dehydrogenases and serum albumins in turtles and various diapsids. Higgins and Rand (1974, 1975) showed that the serum proteins and hemoglobins of *Amblyrhynchus* and *Conolophus* were more similar to each other than to those of *Iguana*. Unfortunately, other iguanines were not examined. Wyles and Sarich (1983) performed immunological comparisons of the serum albumins of 10 species of iguanines including representatives of all eight genera. However, antisera were prepared to the albumins of only four of the species, and comparisons with all others are given only for the antisera to the albumins of *Amblyrhynchus* and *Conolophus*. Because of the incompleteness of the data, only very general phylogenetic inferences can be drawn from them.

The unique colon of iguanines was studied by Iverson (1980, 1982), who reported that the iguanine colon differed from that of all other iguanids and most other lizards in the possession of transverse valves or folds. However, Iverson (1980) felt that the variation in these structures within iguanines was of little value for inferring phylogenetic relationships.

Peterson (1984) has recently surveyed the scale surface microstructure of iguanids. Although some intergeneric variation in the morphology of the scale surface is known to occur in iguanines, representatives of only three iguanine genera (*Iguana*, *Dipsosaurus*, and *Sauromalus*) have been studied at this time.

One final hypothesis about iguanine relationships deserves mention. At the prompting of a colleague (Ernest Williams) some twenty-five years ago, Richard Etheridge drew up a phylogenetic diagram depicting his views on the interrelationships among the iguanid genera. The character basis for this diagram was not specified, and Etheridge (pers. comm., 1981) informs me that the relationships shown among the iguanine genera were strongly influenced by his knowledge about the geographic distributions of these animals. Although he never intended the diagram to be published, it has been published in modified form (Paull et al., 1976; Peterson, 1984), and has also appeared in several graduate theses. I reproduce the original diagram here (Fig. 4), noting that its creator does not grant the hypothesis the conviction seemingly implied by a branching diagram.

GOALS OF THIS STUDY

A detailed study aimed at revealing the pattern of phylogenetic relationships among the various iguanine lizards is sorely needed. It would provide invaluable information for the many people studying other aspects of iguanine biology, particularly in an evolutionary context. I have attempted such a study here with the following as my goals: (1) to provide

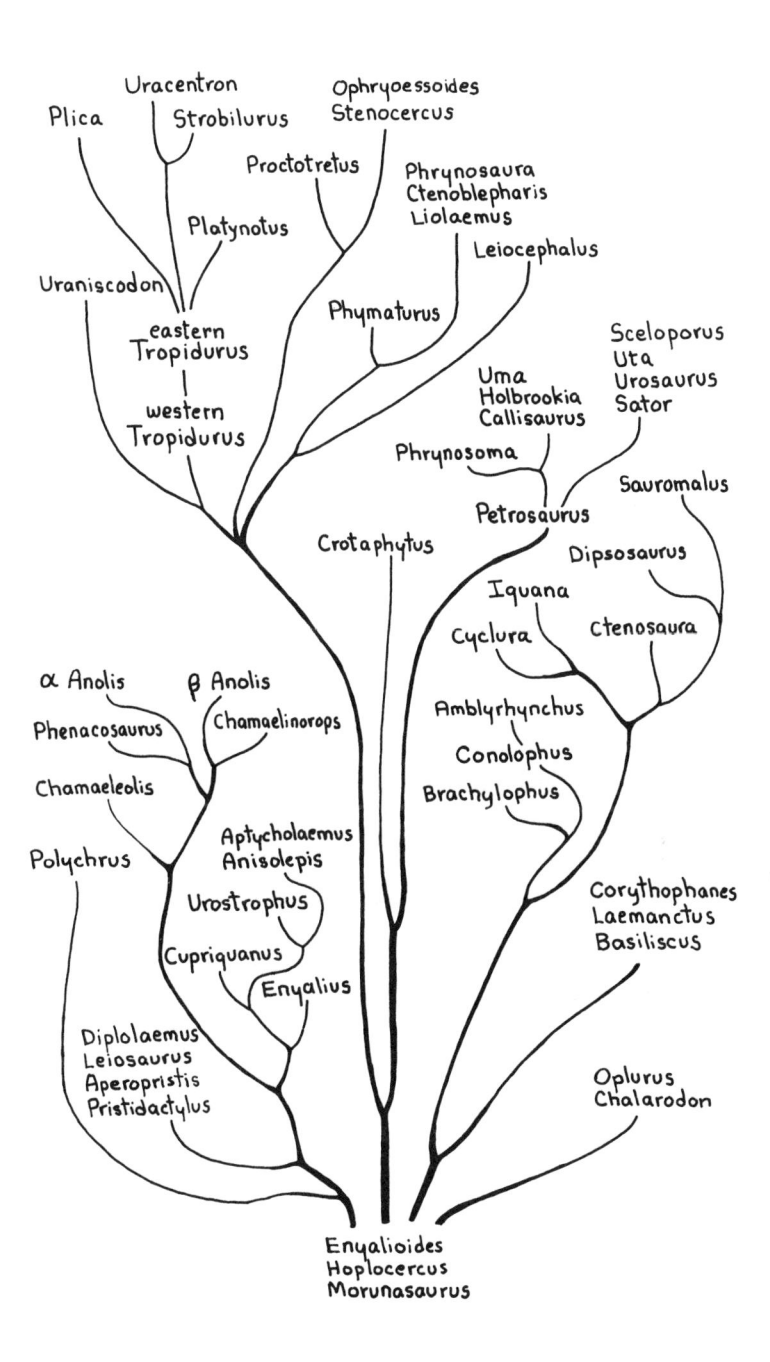

FIG. 4. Etheridge's phylogeny of the Iguanidae.

a diagnosis and a description of the group, including the evidence for its monophyletic status; (2) to describe variation in the iguanine skeleton, as well as review certain aspects of nonskeletal morphology; (3) to generalize characters based on this variation; (4) to assess the polarity of these characters and, thus, their usefulness as evidence for close phylogenetic relationship; (5) to determine the phylogenetic relationships that most reasonably account for the distributions of these characters among various iguanine lizards; and (6) to provide diagnoses and other pertinent information for monophyletic groups within iguanines. The pursuit of these goals has raised numerous questions, some of which are also addressed.

MATERIALS AND METHODS

SPECIMENS

The great majority of the specimens examined in this study were partially intact, complete skeletons that had been prepared by hand or with dermestid beetles. Many skulls prepared by the above methods and some disarticulated skeletons were also examined. Additionally, I have studied radiographs of wet specimens and some whole wet specimens. Observations were made with the naked eye or with the aid of a binocular dissecting microscope. Drawings were made using the camera lucida on a Wild binocular dissecting microscope, from tracings of photographs and radiographs, and freehand. The specimens examined are listed in Appendix I.

PHYLOGENETIC ANALYSIS

I have used the phylogenetic method elaborated by Hennig (1966), and subsequently by many others, to study the relationships of iguanine lizards. Reviews of this method can be found in Eldredge and Cracraft (1980) and Wiley (1981). Hennig's method was formulated specifically for the purpose of evaluating the monophyletic status of groups of organisms, and he stressed that only synapomorphies, shared features that have arisen within a group, logically provide evidence for the existence of its monophyletic subgroups (clades). The assessment of relative apomorphy (polarity) is thus crucial to phylogenetic analysis. I have used the method of outgroup comparison (Watrous and Wheeler, 1981; Farris, 1982; Maddison et al., 1984) for determining character polarity. Although early fossil iguanids might be used as outgroups, I have avoided this practice until their relationships to extant forms are better understood. When ontogenetic transformations are adequately known, I have used the transformations themselves as characters (de Queiroz, 1985).

Outgroups used in this study were the members of four suprageneric groups of iguanids thought to be outside of the iguanine clade (Etheridge and de Queiroz, 1988): basiliscines, crotaphytines, morunasaurs, and oplurines. Maddison et al. (1984) have shown that the precise pattern of relationships among the ingroup and various outgroups has a profound influence on the assessment of plesiomorphy and apomorphy. Unfortunately, relationships among the major groups of iguanids are poorly understood at present (Etheridge and de Queiroz, 1988). Therefore, I selected my four outgroups as much for convenience as for any notion that they were closely related to iguanines. Each of the outgroups is relatively small, which enabled me to examine representatives of all of their included genera and most of their included species. Nevertheless, each one of these outgroups has at some time been proposed as the closest relative of iguanines: basiliscines by Etheridge (1964a); crotaphytines by Savage (1958), who actually included them within iguanines, morunasaurs by Etheridge (Fig. 4), who considered them to be part of a group ancestral to various iguanid lineages; and oplurines by Avery and Tanner (1971).

Evidence about character polarity based on outgroups is not always unambiguous, especially in cases such as this one in which relationships of the various outgroups to the ingroup are unclear (Maddison et al., 1984; Donoghue and Cantino, 1984). I made a compromise between using the maximum number of characters and using only those characters whose polarities were completely unambiguous based on the outgroup evidence. When all four outgroups suggested the same polarity, that polarity was accepted. When all members of all four outgroup taxa did not suggest the same polarity, I considered the polarity determinable only in those cases where all members of three of the four outgroups suggested the same polarity. In all other cases I considered the polarity undeterminable. The reasoning behind this decision is given in Appendix II.

BASIC TAXA

I chose the iguanine genera recognized by Etheridge (1982) as the basic taxa among which phylogenetic relationships were to be determined. Ideally, the basic taxa would all be monophyletic; however, this information is currently unavailable for the iguanine genera. Although it would be preferable from a theoretical viewpoint to use less inclusive taxa (e.g., species) as basic taxa and then determine which genera are monophyletic, this alternative has practical limitations. The characters used in this study are primarily osteological, and many iguanine species are either poorly or not at all represented by osteological preparations. Furthermore, variation within iguanine genera generally appears to be much less than variation among them. For these reasons, using genera rather than species as basic taxa probably gives a more accurate estimation of the range of variation within basic taxa. Nevertheless, overall similarity should not be taken as evidence for monophyly, and I attempt here not only to determine relationships among the iguanine genera but also to evaluate the evidence for the monophyletic status of each. These are not necessarily independent issues.

THE PROBLEM OF VARIATION

Character variation within basic taxa is a problem seldom addressed in the systematic literature. Contrary to the impression given in many systematic studies, such variation is ubiquitous, especially when the basic taxa are higher taxa that are themselves composed of diverse subgroups. Nevertheless, many systematic studies apparently ignore this variation or are based on such small samples that no variation within basic taxa is detected. Many of the characters used in this study vary within one or more of the iguanine genera, the basic taxa of this analysis. To ignore these characters would be to throw away information; for although they contain ambiguities, they also suggest relationships among certain taxa. One very important reason for retaining characters that vary within basic taxa is that these are the only characters that can reveal the paraphyletic status of a basic taxon. I have thus retained certain variable characters in my analysis. However, in cases where the variation within basic taxa was so great that it obscured any pattern of variation among them, the character was eliminated. Because characters form a continuum from those that vary within all basic taxa to those that vary only among them, the decision as to which characters were too variable to be useful at this level of analysis was necessarily arbitrary.

Variation within basic taxa is not itself a problem. For example, variation within basic taxa involving different characters from those that vary among them is simply irrelevant to an analysis of the relationships among these taxa. Even when variation within and among basic taxa involves the same characters, the situation is not necessarily problematic. If the variation within basic taxa characterizes all recognizable subgroups that are units of phylogenetic ancestry and descent (e.g., evolutionary species), or if the basic taxa themselves are such units, then the variable presence of a derived character in two or more basic taxa may be attributable to a polymorphic ancestral population. In contrast, when variation within basic taxa occurs among one or more of their monophyletic subgroups, this variation represents homoplasy. Because homoplastic variation within a basic taxon makes an assessment of its ancestral condition ambiguous, such variation is problematic. The conclusion that variation within a taxon represents homoplastic similarity with a condition occurring in another taxon rests on the assumption that the first taxon is monophyletic. This is a critical point. The variable presence of a derived character in a paraphyletic taxon does not necessarily require homoplasy. In fact, the very characters that give evidence of a taxon's paraphyletic status necessarily vary within that taxon. For this reason, the monophyletic status of basic taxa should be continually reevaluated in light of variation within them.

Variation within basic taxa that is assumed to represent homoplasy can be handled in various ways, none of which is without drawbacks. The most appropriate method for a particular case will depend on the amount and nature of the variation, as well as on the assumptions that the investigator is reasonably able to make. I discuss below four methods for handling variation within basic taxa.

1. One means of handling variation within basic taxa is to abandon these taxa in favor of less inclusive basic taxa.: for example, one might use species instead of genera. Unfortunately, this practice carries no guarantee that variation will not exist within the new basic taxa. A practical disadvantage of this method is that it necessarily increases the number of basic taxa. Furthermore, the number of specimens may be distributed in such a way that choosing the more inclusive groups as basic taxa gives a more accurate picture of the range of variation. A lack of variation within less inclusive taxa may result simply from small samples.

2. Kluge and Farris (1969) handled variation within basic taxa by subdividing each taxon into two taxa, each coded invariable for one of the alternative states of the variable character. This method makes no assumptions about phylogenetic relationships other than those involved in the delimitation of the basic taxa, and it should be satisfactory as long as there are few variable characters. As the number of variable characters increases for a given taxon, however, the number of new "taxa" created by subdivision increases geometrically. Therefore, this method will be impractical if any of the basic taxa exhibit more than a few variable characters.

3. Information about relationships within a basic taxon can be used to assess the ancestral condition of a character that varies within the taxon. The ancestral condition can then be assigned to the taxon because it follows that the alternative condition bears homoplastic rather than synapomorphic resemblance to similar conditions in other basic taxa. An obvious drawback of this method is that it requires information about relationships within basic taxa, information that is not always available. An advantage of this method is that it does not increase the number of basic taxa.

4. A fourth means of dealing with variation within basic taxa is to arbitrarily choose one of the alternative conditions as the ancestral condition for the variable taxon. Although it might intuitively seem that the condition that appears to be ancestral on morphological grounds (i.e., the condition that is found in outgroups) is the ancestral state, this conclusion has a hidden bias. Given that the variation represents homoplasy, assigning the morphologically "ancestral" state to the taxon amounts to asserting that the homoplasy is a convergence; assigning the morphologically "derived" state to the taxon amounts to asserting that the homoplasy is a reversal. Furthermore, the most reasonable interpretation of character transformation after construction of a cladogram or phylogenetic tree may conflict with the original choice of an ancestral condition for a variable basic taxon. Therefore, the level at which characters that vary within basic taxa are considered to be synapomorphies should always be reevaluated after an initial analysis that does not take the variation into consideration.

I have already presented reasons for not using less inclusive taxa than the iguanine genera as my basic taxa. In this study, character variation within basic taxa was high enough that the method of Kluge and Farris (1969) would have been impractical. The third method for handling variation within basic taxa could not be used since not enough is known about relationships within iguanine genera to use this information to assess the ancestral condition of characters that vary within them. Therefore, I have been forced to use the last, and perhaps the least satisfactory, means of dealing with variation within basic taxa.

CONSTRUCTION OF BRANCHING DIAGRAMS

The branching diagrams presented in this study are all intended to be cladograms rather than phylogenetic trees (Nelson, cited in Wiley, 1979); in other words, no attempt is made to identify direct ancestors. Cladograms were constructed without the aid of computer programs that minimize the number of instances in which shared characters appearing to be synapomorphic on morphological grounds have to be interpreted as homoplastic in light of other characters. In most cases, these cladograms were checked using the Wagner programs in Farris's PHYSYS computer package installed on the California State University CYBER system; however, in the analysis of relationships within *Ctenosaura*, character incongruence was judged to be so low that such a check was unnecessary. In no case did the computer analysis find a cladogram with fewer homoplasies than I was able to find without it.

IGUANINE MONOPHYLY

The family Iguanidae is a heterogeneous assemblage of iguanian lizards distinguished from other iguanians (agamids and chamaeleontids) primarily by their possession of pleurodont rather than acrodont teeth (e.g., Boulenger, 1884; Cope, 1900; Camp, 1923). Pleurodonty occurs in nearly all other squamates (except trogonophid amphisbaenians) and in most other lepidosauromorphs (except sphenodontidan rhynchocephalians), although the teeth of early lepidosauromorphs are unlike those of most squamates in that they are set in shallow depressions (Gauthier et al., 1988). Thus, the primary diagnostic feature of Iguanidae is probably plesiomorphic for Iguania and is not, therefore, evidence for the monophyletic status of the family. Other characters given in a recent diagnosis of Iguanidae (Moody, 1980) all either appear to be plesiomorphic for Iguania or do not characterize all iguanids (Estes et al., 1988). There is presently no evidence that Iguanidae is monophyletic. Renous (1979) and Moody (1982) have even suggested that some iguanids may be more closely related to some or all of the acrodont iguanians than they are to other iguanids.

Because Iguanidae is unlikely to be monophyletic, it seems advisable to break this family into groups for which there is evidence of monophyly. A number of presumably monophyletic groups of iguanids have been proposed and given informal names (Savage, 1958; Etheridge, 1964a, 1976 in Paull et al.; Estes and Price, 1973; Etheridge and de Queiroz, 1988); however, it is beyond the scope of this paper to revise the taxonomy of all pleurodont iguanians. Furthermore, the relationships of these informal groups to agamids and chamaeleontids are unclear (Estes et al., 1988). Therefore, I provisionally retain the taxon Iguanidae, emphasizing that it may be paraphyletic and thus may have to be abandoned at some future time.

I present evidence below for the monophyletic status of iguanine iguanids. Because I feel that it would be premature to disband Iguanidae, but because I also feel that the monophyletic groups of iguanians should be recognized, I resurrect the taxon Iguaninae.

Traditionally, recognition of a taxon within a larger one is accompanied by the assignment of all members of the larger taxon to a taxon at the same categorical level as the new one. This often leads to the recognition of some new paraphyletic group, especially in cases such as this one where the author has adequate knowledge of only part of the larger group being subdivided. Because paraphyletic taxa are undesirable in a phylogenetic taxonomy, I chose a logical alternative: to leave the remaining iguanids unassigned to taxa of rank equal to that of Iguaninae. Indeed, following Gauthier et al. (1988), I have more or less abandoned the use of categorical ranks. My use of "genera" as basic taxa is the result of historical inertia, and I do not mean to imply that they are equivalent in any biologically meaningful way. Similarly, the use of standard endings for new taxa proposed in this study is an attempt to conform with nomenclatural rules rather than to designate categorical ranks. Although these practices go against tradition, they convey the current state of knowledge precisely. If taxonomies are to serve as summaries of phylogenetic knowledge, conveying such information is more valuable than being able to pigeonhole all organisms equally.

Iguaninae Bell 1825

Diagnosis. The following combination of characters is diagnostic for iguanines, separating the members of this group not only from all other iguanids but also from all other iguanians and probably from all other squamates:
1. some caudal vertebrae with two pairs of transverse processes that diverge from one another (Etheridge, 1967);
2. the presence of transverse valves or folds in the colon (Iverson, 1980, 1982);
3. crowns of posterior marginal teeth laterally compressed, anteroposteriorly flared, and often polycuspate (Etheridge, 1964a);
4. the supratemporal lies primarily on the posteromedial surface of the supratemporal process of the parietal;
5. primarily herbivorous diet (H. M. Smith, 1946).

This list does not include all features in which iguanines exhibit a derived condition within Iguanidae; it consists of only those that I consider to be true iguanine synapomorphies evidencing monophyly of the group. The first two are unique to iguanines (within iguanids) and are therefore unproblematic. Each can be used alone to distinguish an iguanine from any other iguanid, though colic anatomy is unknown for many members of the family. The other characters occur in some noniguanine iguanids where I consider them to be convergent, a conclusion necessitated by conflicting distributions of other derived characters. Still other apomorphic characters of iguanines are not included in the diagnosis. These characters are also more widely distributed and may be iguanine synapomorphies (in which case, other convergences will be required) or may serve to diagnose more inclusive clades within the Iguania. These characters are given in the description.

Description. The description below is a list of the apomorphies, plesiomorphies, and characters of uncertain polarity shared by iguanines. In addition, some characters that vary within the group are included. This list is meant to provide a basis for comparison with other iguanids and is intended to be a source of characters that may be useful for examining the relationships of iguanines to the rest of Iguanidae (and Iguania). For this reason, I include only characters that vary within Iguanidae. Morphologies thought to be derived within Iguanidae are indicated by an asterisk (*).

HEAD SKELETON: premaxillary spine exposed dorsally or covered* between nasals; surface of dermatocranium smooth or with irregular rugosities; parietal roof trapezoidal, V-shaped*, or with a posterior midsagittal crest (Y-shaped)*; parietal foramen present, on frontoparietal suture or within frontal*; outlines of osseous labyrinth indistinct (moderately distinct in *Dipsosaurus*); supratemporal situated primarily on posteromedial surface of supratemporal process of parietal*; lacrimal present; postfrontal present; septomaxilla with a large posterodorsal shelf*; epipterygoid present; Meckel's groove closed and fused between anterior end of splenial and mandibular symphysis*; angular present; splenial present; coronoid with large lateral process overlapping dentary; three to eleven premaxillary teeth (most species with a mode of seven); premaxillary teeth without lateral cusps, bicuspid, or tricuspid; crowns of posterior marginal teeth flared, with three to many* cusps; palatine teeth absent*; pterygoid teeth present or absent*; second ceratobranchials long* or short, adpressed or separated* medially; 14 scleral ossicles.

AXIAL SKELETON: vertebral neural spines tall or short*; zygosphenes and zygantra well developed*; mode of 24 or 25* presacral vertebrae; caudal autotomy septa present or absent*; less than 25* to more than 70 caudal vertebrae; some caudal vertebrae with two pairs of transverse processes that diverge from one another*; first rib usually borne on fifth cervical vertebra; usually four pairs of free (cervical) ribs on fifth through eighth presacral vertebrae; usually four pairs of sternal ribs on ninth through twelfth presacral vertebrae; one* or two pairs of xiphisternal ribs attached to 13th and usually 14th presacral vertebrae; all post-thoracic presacral vertebrae usually bear unfused ribs; postxiphisternal inscriptional ribs attached to corresponding bony ribs, with zero to three pairs meeting midventrally to form continuous chevrons.

APPENDICULAR SKELETON: clavicles with or without* flattened lateral shelf; dorsal ends of clavicles articulate with suprascapulae; clavicular fenestrae usually absent (except in *Conolophus*); scapular fenestrae usually present* (variably reduced or absent in *Amblyrhynchus* and *Sauromalus*); posterior coracoid fenestrae present* or absent; lateral arms of interclavicle form angles of 45° to 90° with posterior process; posterior process of interclavicle extending posterior to lateral corners of sternum or not*; sternal fontanelle present but small (approximately equal to interclavicle in width) or absent*; phalangeal formula of manus 2:3:4:5:3; phalangeal formula of pes 2:3:4:5:4.

NONSKELETAL ANATOMY: superciliary scales quadrangular and non-overlapping to elongate and strongly overlapping (Etheridge and de Queiroz, 1988); subocular scales all subequal and quadrangular to one greatly elongate (Etheridge and de Queiroz, 1988); interparietal scale small; transverse gular fold present (weakly developed in *Amblyrhynchus*); pendulous dewlap present* or absent; gular crest present* or absent; middorsal row of enlarged scales present or absent*; subdigital lamellae keeled, without setae; femoral pores present in one or two rows; preanal pores absent; scale surface with honeycomb pattern (Etheridge and de Queiroz, 1988); scale organs lacking elongated spines (Etheridge and de Queiroz, 1988); nasal passage S-shaped, with concha present (Stebbins, 1948); colon with transverse valves or folds (Iverson, 1980).

COMPARATIVE SKELETAL MORPHOLOGY

In this section I describe variation in the iguanine skeleton by region and compare iguanines with basiliscines, crotaphytines, morunasaurs, and oplurines. These descriptions emphasize variation that is relevant to an analysis of relationships among iguanine genera and are not intended to be exhaustive. For more detailed descriptions of the head skeleton that include features common to all iguanines I refer the reader to Oelrich (1956), whose terminology is followed below.

SKULL ROOF

The iguanine skull roof, or the superficial dermatocranial elements of the skull proper, consists of the full complement of bones thought to be plesiomorphic for squamates. Other iguanians may lack some of these bones, most commonly lacrimals and postfrontals. The quadrate, a splanchnocranial bone, is also described in this section. Bones of the iguanine skull roof, palate, braincase, and lower jaw are illustrated in Figures 5 and 6.

Premaxilla (Figs. 5A, 6A, 7, 8). The premaxilla is the anteriormost bone in the skull. Postembryonically, it is a median, unpaired bone that is sutured laterally with the maxillae, posterodorsally with the nasals, and posteroventrally with the vomers. The premaxilla bears a ventrally directed incisive process near its posteroventral end on the midline (Fig. 7). Foramina for the maxillary arteries (Oelrich, 1956) penetrate the premaxilla on either side of the incisive process. In most iguanines (*Brachylophus, Cyclura, Ctenosaura, Iguana, Dipsosaurus,* and *Sauromalus*) and all outgroups examined, the ventral surface of the premaxilla bears well-developed posteroventral extensions where it sutures with the maxillae. A weak ventral crest runs along the posterior edge of the premaxilla from its posterolateral corners to the base of the incisive process. This crest is generally not pierced by the foramina for the maxillary arteries. In *Amblyrhynchus*, the posterolateral extensions of the premaxilla are small and the posterolateral corners of this bone are concomitantly closer to the base of the incisive process (Fig.7A) than are those of other iguanines (Fig. 7B). *Conolophus* differs from the typical iguanine pattern in having large posteroventral crests that continue up the sides of the incisive process and are pierced or notched by the foramina for the maxillary arteries (Fig. 7C). In *Sauromalus slevini* these foramina also pierce the ventral crest of the premaxilla; however, the condition does not appear to be homologous with that seen in *Conolophus*. In *Conolophus* the crests are pierced because of their enlargement, while in *S. slevini* the foramina appear to have moved posteriorly.

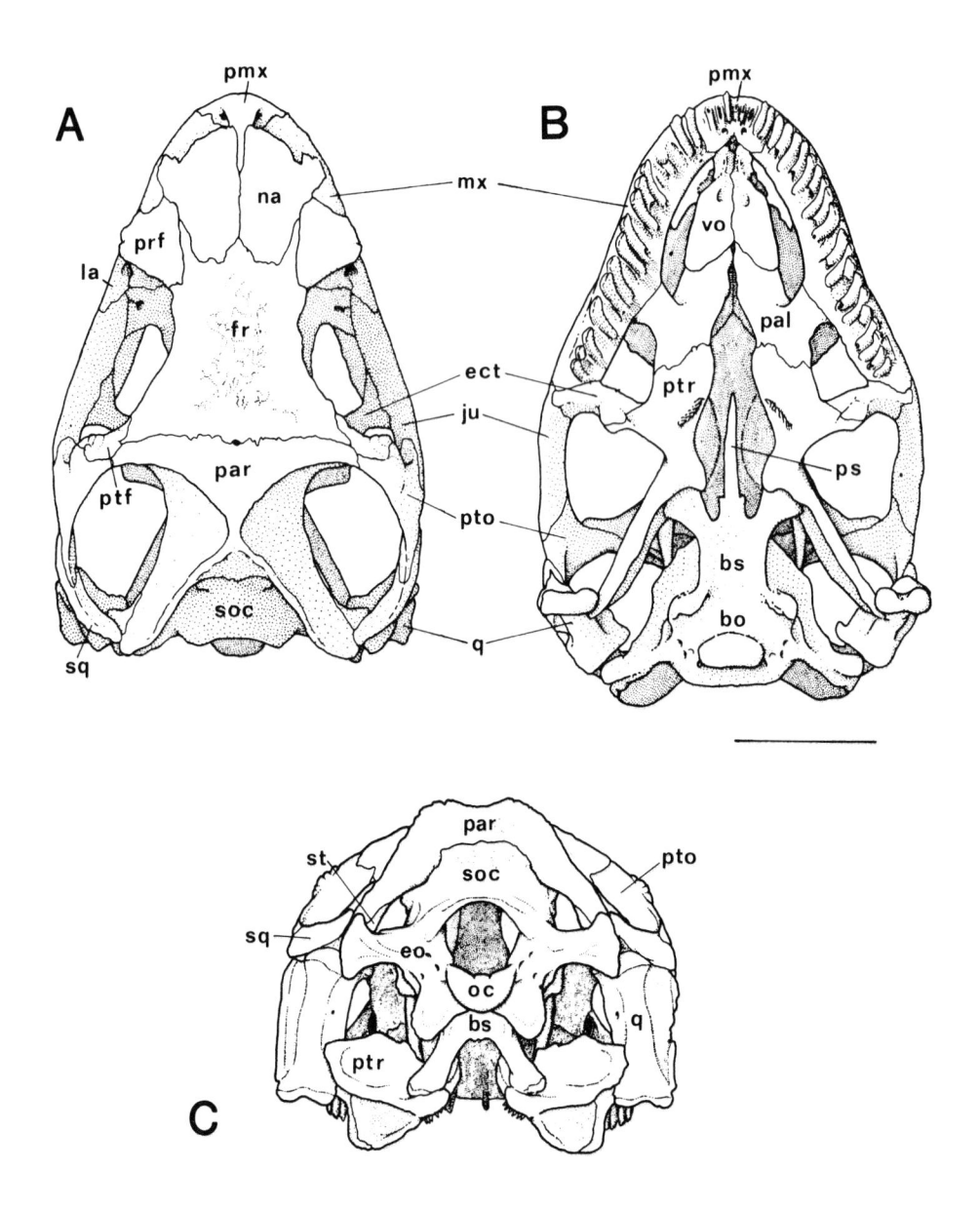

FIG. 5. Skull of *Brachylophus vitiensis* (MCZ 160254): (A) dorsal view, (B) ventral view, (C) posterior view. Scale equals 1 cm. Abbreviations: bo, basioccipital; bs, parabasisphenoid; ect, ectopterygoid; eo, exoccipital-opisthotic; fr, frontal; ju, jugal; la, lacrimal; mx, maxilla; na, nasal; oc, occipital condyle; pal, palatine; par, parietal; pmx, premaxilla; prf, prefrontal; ps, parasphenoid rostrum; ptf, postfrontal; pto, postorbital; ptr, pterygoid; q, quadrate; soc, supraoccipital; sq, squamosal; st, supratemporal; vo, vomer.

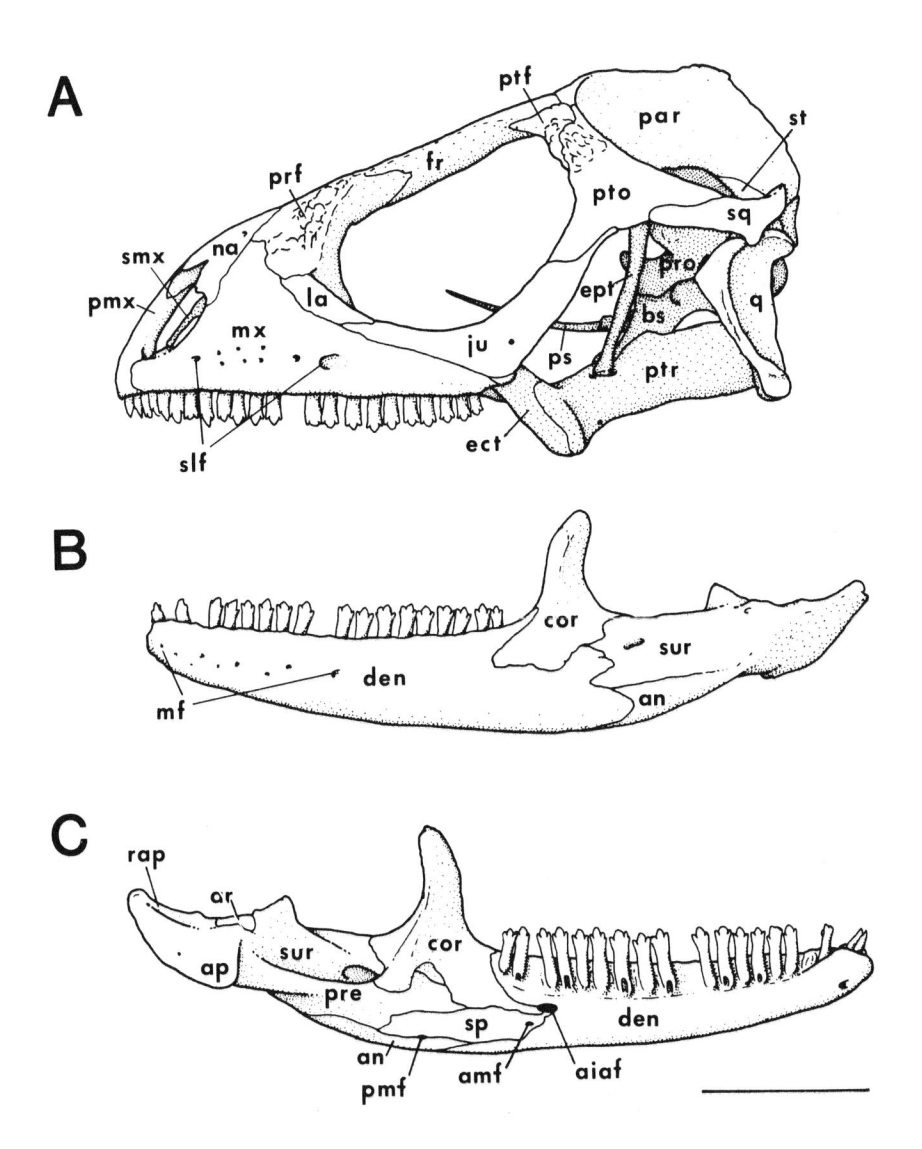

FIG. 6. Skull and mandible of *Brachylophus vitiensis* (MCZ 160254): (A) lateral view of skull, (B) lateral view of left mandible, (C) lingual view of left mandible. Scale equals 1 cm. Abbreviations: aiaf, anterior inferior alveolar foramen; amf, anterior mylohyoid foramen; an, angular; ap, angular process; ar, articular; bs, parabasisphenoid; cor, coronoid; den, dentary; ect, ectopterygoid; ept, epipterygoid; fr, frontal; ju, jugal; la, lacrimal; mf, mental foramina; mx, maxilla; na, nasal; par, parietal; pmf, posterior mylohyoid foramen; pmx, premaxilla; pre, prearticular; prf, prefrontal; pro, prootic; ptf, postfrontal; pto, postorbital; ptr, pterygoid; ps, parasphenoid rostrum; q, quadrate; rap, retroarticular process; slf, supralabial foramina; smx, septomaxilla; sp, splenial; sq, squamosal; st, supratemporal; sur, surangular.

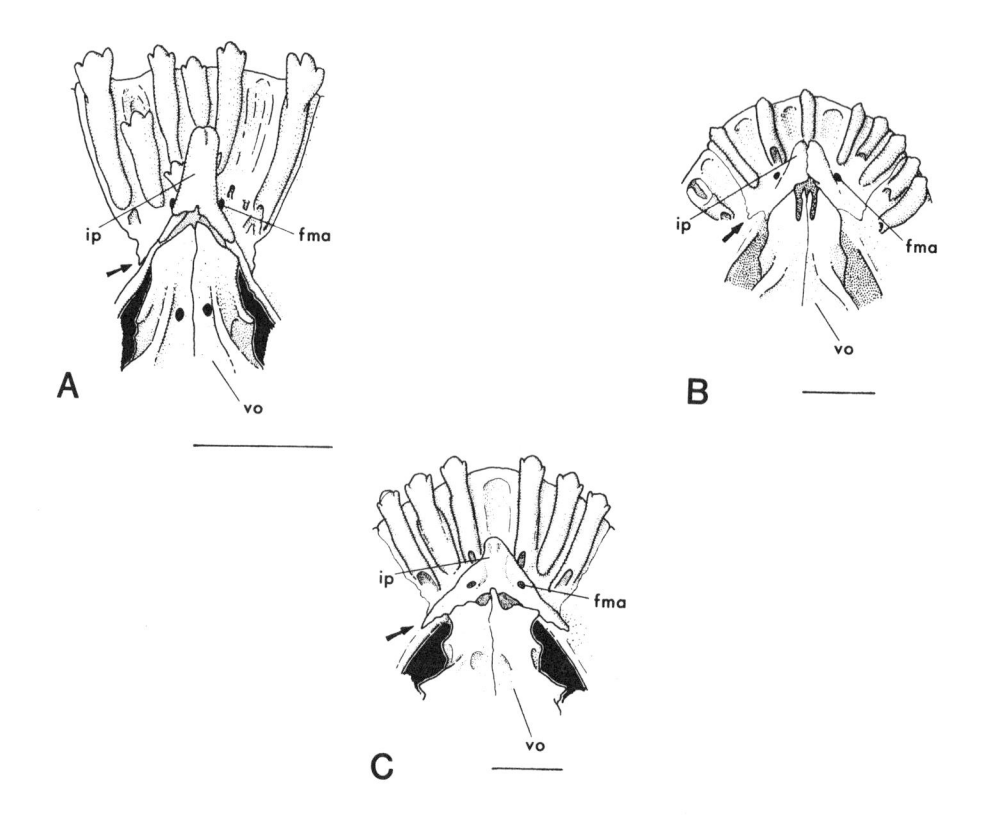

FIG. 7. Posteroventral views of the premaxillae of (A) *Amblyrhynchus cristatus* (RE 1396), (B) *Cyclura cornuta* (RE 383), and (C) *Conolophus pallidus* (RE 439), showing the small posterolateral processes (arrows) of the premaxilla in *Amblyrhynchus* and the large lateral crests of the incisive process that are pierced by foramina for the maxillary arteries in *Conolophus*. Scale equals 0.5 cm. Abbreviations: fma, foramen for maxillary artery; ip, incisive process; vo, vomer.

The premaxilla of *Amblyrhynchus* differs from that of basiliscines, crotaphytines, morunasaurs, oplurines, and all other iguanines in several other ways. Its anterior edge forms a nearly flat plate rather than being arched, and its nasal process is nearly vertical instead of sloping backwards.

The nasal process of the premaxilla, or the premaxillary spine, meets the nasals posterodorsally (Figs. 5A, 6A, 8). In cross section this process is roughly triangular, with the apex of the triangle pointing posteroventrally. The shape of this triangle varies considerably, ranging from broad-based and low in *Ctenosaura* and *Conolophus* to narrow-based in *Amblyrhynchus*, *Dipsosaurus*, and *Sauromalus* to nearly oval in *Cyclura cornuta*. Differences between morphological extremes are great, but the extremes grade more or less continuously into one another through intermediates; thus, there are no gaps to

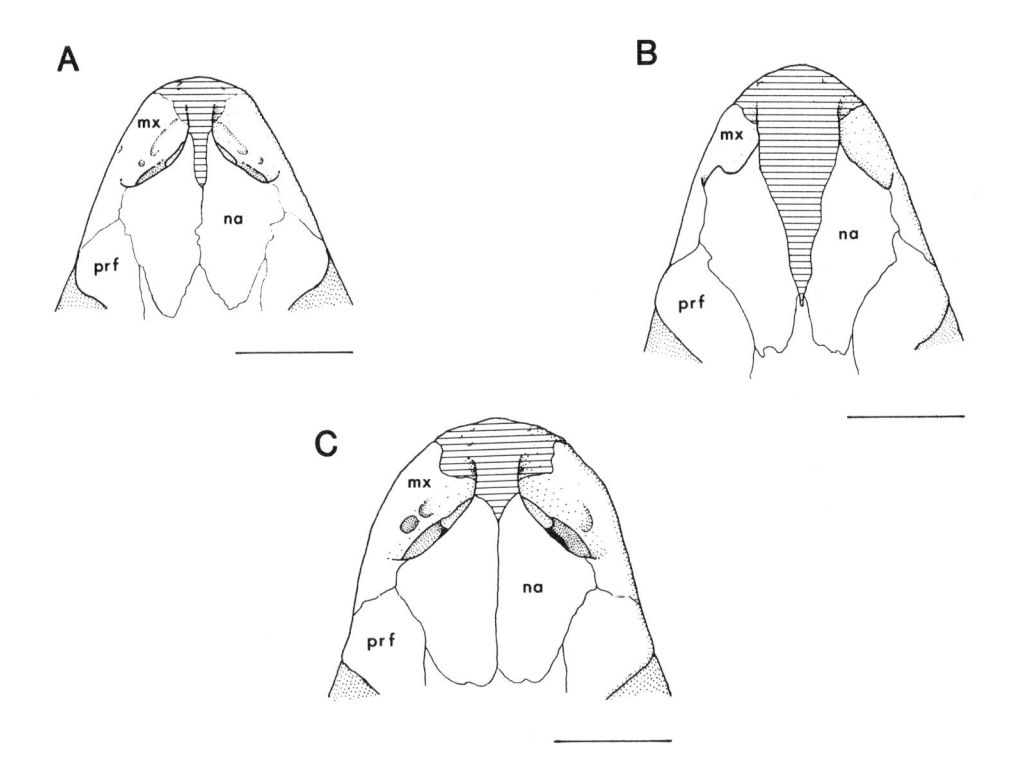

FIG. 8. Dorsal views of the preorbital portions of the skulls of (A) *Sauromalus varius* (RE 308), (B) *Ctenosaura hemilopha* (RE 1964), and (C) *Conolophus subcristatus* (MVZ 77314), showing differences in the degree to which the nasal process of the premaxilla is covered dorsally by the nasals. Scale equals 1.0 cm. Premaxilla is shaded. Abbreviations: mx, maxilla; na, nasal; prf, prefrontal.

separate discrete character states. Furthermore, variation within *Cyclura* is greater than that between many of the genera. For these reasons I have chosen not to use variation in the cross-sectional shape of the premaxillary spine as a character in phylogenetic analysis.

When viewed dorsally, the posterior exposure of the nasal process is variable within iguanines (Fig. 8). In *Brachylophus, Ctenosaura, Cyclura, Dipsosaurus, Iguana*, and *Sauromalus*, the nasal process of the premaxilla is exposed dorsally or covered only slightly dorsolaterally by the nasals. Although differences in the extent of this overlap and the length of the premaxillary spine yield strikingly different morphologies (Fig. 8A, B), intraspecific variation in these features is too great to permit their subdivision into character states. In *Amblyrhynchus* and *Conolophus*, however, one finds a condition not seen in any other iguanine. The nasals of these genera cover the premaxillary spine posteriorly so that the posteriormost portion of the spine that is visible dorsally falls short of the transverse plane at the posterior ends of the fenestrae exonarinae (bony external nares or nasal fossae). In *Cyclura cornuta*, *C. pingius*, and *Iguana delicatissima*, the visible portion of the nasal process of the premaxilla also fails to reach this plane; however, this condition results from enlargement of the nasal fossae in these species rather than from increased overlap of the premaxillary spine by the nasals.

Basiliscines, oplurines, and morunasaurs generally have the nasal process of the premaxilla exposed dorsally, at least along the midline. In crotaphytines, more extensive overlap of the premaxillary spine by the nasals may occur, but never to the extent seen in *Amblyrhynchus* and *Conolophus*.

Nasals (Figs. 5A, 6A, 8). This pair of bones lies along the midline of the skull roof immediately posterior to the fenestrae exonarinae, forming the roof of the nasal capsule. The relative size of the nasals varies considerably among iguanines. *Brachylophus*, *Conolophus*, *Ctenosaura*, most *Cyclura*, *Dipsosaurus*, *Iguana iguana*, and *Sauromalus* exhibit a condition that is widespread outside of the iguanines in which the nasals are of moderate size. In *Conolophus* the nasals are subequal to the frontals in length, and in the remaining taxa the nasals are shorter than the frontals. The nasals of *Amblyrhynchus* are greatly enlarged, accompanying the enlargement of the entire nasal capsule. The increased size of the nasal capsule probably enables it to house the large nasal salt glands of this species (figured by Dunson, 1969, 1976). Because enlargement of the nasals is one component of enlargement of the nasal capsule, I did not treat the two as separate characters.

The relative size of the nasals is also correlated with the size of the fenestrae exonarinae (Fig. 9). In some species of *Cyclura*, most notably *C. cornuta* (Fig. 9A) and *C. pinguis*, and in *Iguana delicatissima*, enlargement of the fenestrae exonarinae results from emargination at the anterior edges of the nasals, reducing the relative size of these bones. Because this apomorphic condition occurs only in some *Iguana* and *Cyclura*, I have not used it for analyzing intergeneric relationships. The feature may, however, provide useful information for uncovering relationships within these genera.

Septomaxillae (Fig. 6A). The septomaxillae are thin, paired bones lying in the anterior portion of the nasal capsule on either side of the nasal septum. These bones rest atop the vomers and, except in *Iguana delicatissima*, they contact the roof of the nasal capsule (premaxilla or nasals) dorsally. In most iguanines, each septomaxilla is relatively flat on its dorsal surface, though a low ridge often appears near the lateral edge of the bone. Unlike all other iguanines, the septomaxillae of *Amblyrhynchus* bear sharp longitudinal crests on their anterodorsal surfaces. These crests appear to be apomorphic, since they are absent in all four outgroups. The septomaxillae of crotaphytines and oplurines are similar to those of iguanines in that they each send a long shelf posterodorsally. Those of morunasaurs and *Basiliscus* are relatively small and are folded to form a transverse ridge. The septomaxillae of *Laemanctus* are very small, and those of *Corytophanes* are possibly absent. A dorsal septomaxillary crest is seen in some morunasaurs, but this crest is entirely different from that of *Amblyrhynchus*, extending transversely rather than longitudinally.

Prefrontals (Figs. 5A, 6A, 8, 9). The prefrontals have roughly the shape of a triangular pyramid whose apex forms the posterolateral corner of the nasal capsule.

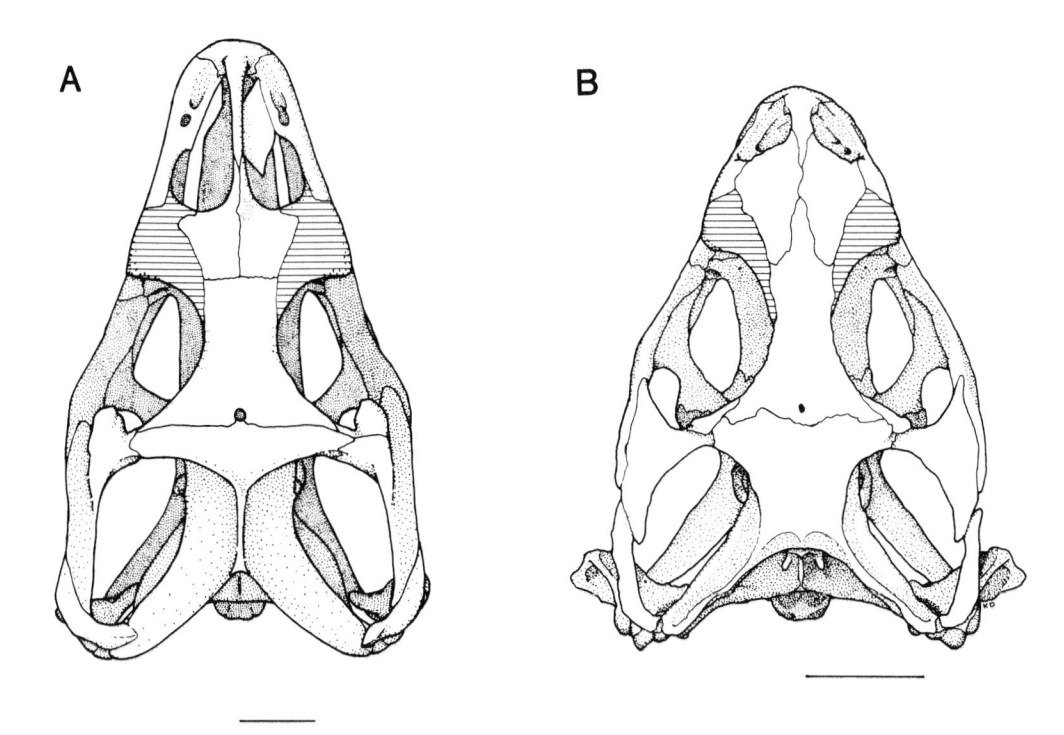

FIG. 9. Dorsal views of the skulls of (A) *Cyclura cornuta* (AMNH 57878) and (B) *Sauromalus obesus* (RE 467), showing differences in the relative sizes of the external nares and the presence or absence of prefrontal contribution to the posterior margin of the nares. Prefrontal is shaded. Scale equals 1 cm.

Boulenger (1890) noted that the contribution of this bone to the posterior margin of the fenestra exonarina (Fig. 9A) distinguished *Metopoceros cornutus* (*Cyclura cornuta*) from all other iguanines. In the other iguanines, maxilla and nasal meet anterior to the prefrontal and exclude it from the margin of the fenestra (Fig. 9B). This condition is undoubtedly plesiomorphic, since it is found in all basiliscines, crotaphytines, morunasaurs, and oplurines, and in most other lizards (except varanoids, *Shinisaurus*, and chamaeleons). In *C. cornuta*, contribution of the prefrontal to the margin of the fenestra exonarina is related to enlargement of the latter. Although the prefrontals of other species with large bony external nares do not enter the fenestra margin, none of these have bony external nares as large as those of *C. cornuta*. Because the apomorphic condition is found in only one species of a genus containing eight, it provides no information about intergeneric relationships.

A large lacrimal foramen, bounded by the prefrontal medially and the lacrimal laterally, pierces the anterior wall of the orbit (Fig. 10). Just posterior to the lacrimal foramen four bones-lacrimal, jugal, palatine, and prefrontal-approach one another, and one of two

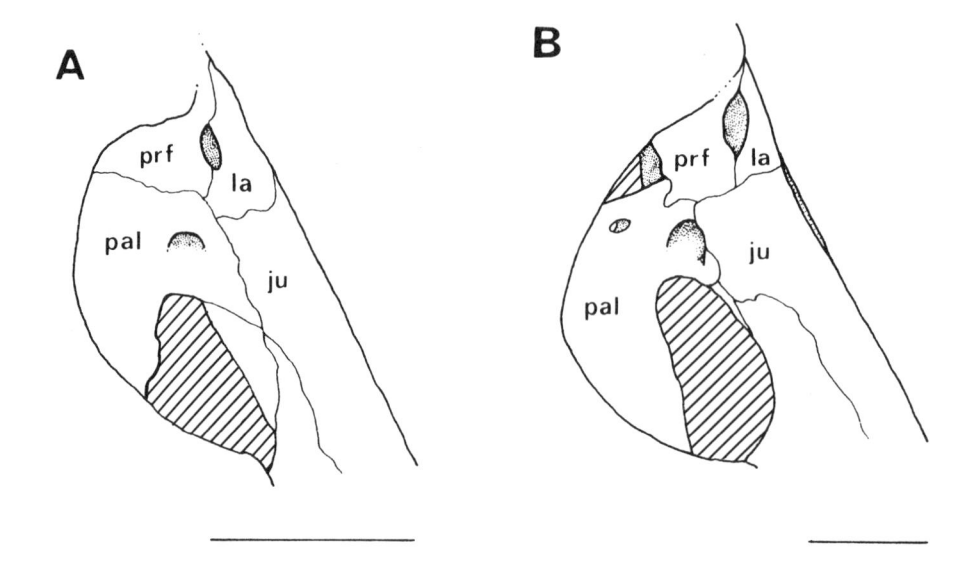

FIG. 10. Posterodorsal views of the anterior orbital regions of (A) *Brachylophus fasciatus* (RE 1888) and (B) *Conolophus pallidus* (MCZ 79772), showing differences in the contacts of the bones in this region. In A the lacrimal (la) contacts the palatine (pal); in B the prefrontal (prf) contacts the jugal (ju). Scale equals 0.5 cm.

different contacts may be established: between lacrimal and palatine (Fig. 10A), or between prefrontal and jugal (Fig. 10B). Occasionally all four bones meet at a single point. Because the lacrimal-palatine contact occurs in all outgroups except crotaphytines, this condition appears to be plesiomorphic for iguanines. The apomorphic prefrontal-jugal contact is the common condition in *Amblyrhynchus* and *Conolophus*. The plesiomorphic condition is characteristic of most species of all other iguanine genera, although *Ctenosaura clarki*, *Cyclura carinata*, *C. cornuta*, and *C. ricordii* appear to be characterized by the apomorphic condition. This character is also variable within some iguanine species and occasionally even varies between right and left sides of a specimen thus decreasing its value as a systematic character.

Frontal (Figs. 5A, 6A, 11, 12). The frontal of postembryonic iguanines is an unpaired median bone that forms the dorsal borders of the orbits. It is sutured anteriorly with the nasals and anterolaterally with the prefrontals. Posteriorly the frontal meets the parietal, forming a transverse suture, and posterolaterally it contacts the postfrontal and variably the postorbital bones. The proportions of the frontal vary within Iguaninae. Generally, the frontal is about as wide (at the frontoparietal suture) as it is long (maximum exposed length dorsally), or is longer than wide. Four iguanines have frontals that are markedly wider than long: *Amblyrhynchus* (Fig. 11) has the relatively widest frontal of all iguanines,

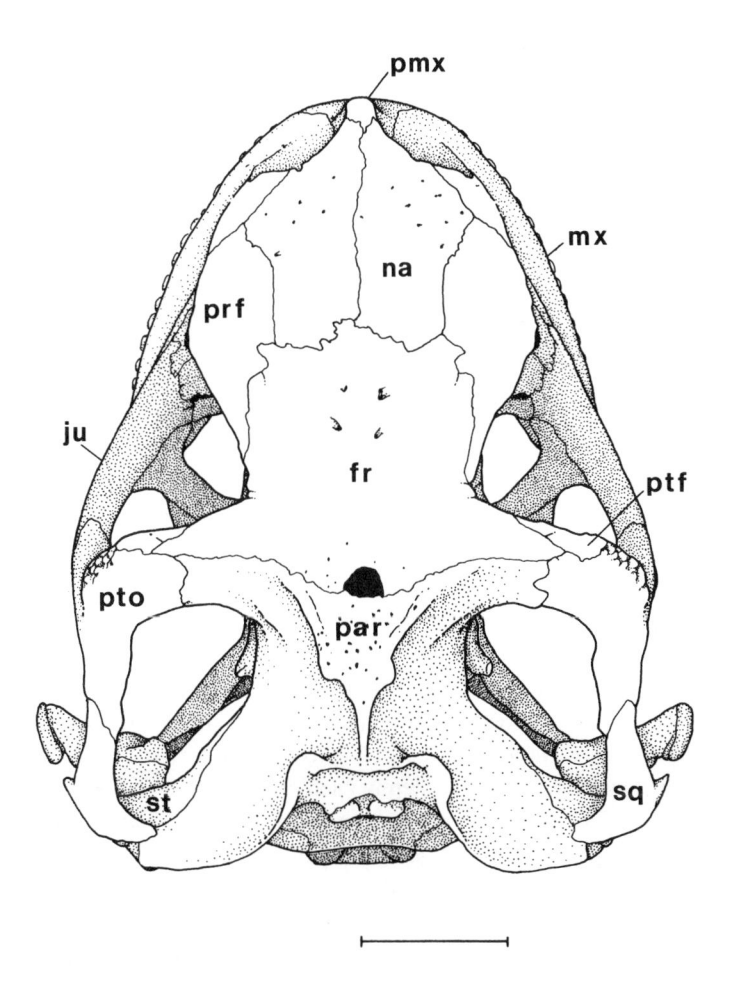

FIG. 11. Dorsal view of the skull of *Amblyrhynchus cristatus* (MVZ 67721), showing the short, wide frontal and the wedge-shaped orbital borders. Scale equals 1 cm. Abbreviations: fr, frontal; ju, jugal; mx, maxilla; na, nasal; par, parietal; pmx, premaxilla; prf, prefrontal; ptf, postfrontal; pto, postorbital; sq, squamosal; st, supratemporal.

while *Conolophus*, *Cyclura cornuta*, and *Iguana delicatissima* are intermediate between *Amblyrhynchus* and the other iguanines. In *C. cornuta* and *I. delicatissima* reduction in frontal length may be related to enlargement of the fenestrae exonarinae, although *Cyclura pinguis*, which has large bony external nares, has a frontal that is about as wide as it is long.

The width of the frontal between the orbits exhibits substantial intergeneric variation in iguanines, but the pattern of variation is obscured by correlation of this feature with size. Small species and juveniles of large species have relatively large orbits and concomitantly narrow frontals. As members of large species grow, relative orbit size decreases and relative frontal width increases correspondingly. Nonetheless, *Brachylophus* has seemingly wider frontals than would be predicted on the basis of its body size and knowledge of frontal allometry in other iguanines. A detailed study of this allometry was not undertaken, and I did not use variation in frontal width as a systematic character.

Along with the prefrontals and postfrontals, the frontal forms the dorsal borders of the orbits, which have a unique shape in *Amblyrhynchus*. In this genus, the dorsal orbital borders are wedge-shaped, with the apices of the wedges pointing medially when viewed from above (Fig. 11). In all other iguanines and all outgroups examined, the dorsal borders of the orbits are more or less smoothly curved (Figs. 5A, 9), though the precise shape varies among taxa.

Dipsosaurus has a pair of relatively large openings at or near the suture between frontal and nasals, one on each side of the midline. No other iguanine nor any outgroup that I have examined has these, although some have much smaller foramina in roughly the same location that may or may not be homologous. An area just anterior to the frontal on the midline fails to ossify in certain other iguanids (e.g., *Corytophanes* and morunasaurs), but it is not divided into paired openings as in *Dipsosaurus*.

On the anteroventral surface of the iguanine frontal at the posterior end of the nasal capsule, one or two pairs of crests may develop (Fig. 12). The lateral pair are the cristae cranii, which form the dorsolateral walls of the olfactory tract and are invariably present. In most iguanines these cristae extend continuously from the frontal onto the prefrontals (Fig. 12A); however, in *Conolophus* and in *Ctenosaura defensor*, the frontal portions of the cristae project anteriorly, forming a step in each crista between its frontal and prefrontal portions (Fig. 12B). Basiliscines, crotaphytines, morunasaurs, and oplurines all possess cristae cranii resembling those in the majority of iguanines.

A second pair of crests, located medial to the cristae cranii, is variably developed in iguanines. These crests extend posterolaterally from the anterior end of the frontal at the midline towards the cristae cranii. This medial pair of crests is absent in most iguanines (Fig. 12A). Weakly developed medial crests are seen in *Brachylophus* and some *Ctenosaura*, and somewhat larger ones occur in *Conolophus* (Fig. 12B). *Amblyrhynchus* exhibits the strongest development of the medial crests and also undergoes considerable ontogenetic change in this feature. At hatching, the medial crests of *Amblyrhynchus* are similar to those of other iguanines that possess the crests, but are more strongly developed and are united anteriorly to form a single median crest (Fig. 12C). During postembryonic development the median portion elongates and grows ventrally, while the posterior ends of the crests also grow ventrally and become continuous with the cristae cranii. The end result is the formation of a pair of deep pockets separated by a median crest on the ventral surface of the frontal bone (Fig. 12D). Large nasal salt glands (Dunson, 1969, 1976) probably fill these pockets in entire specimens.

Continuous morphological variation and a high degree of intergeneric overlap in the range of this variation limit the usefulness of the medial pair of frontal cristae as characters in phylogenetic analysis. The unique condition seen in *Amblyrhynchus*, however, is easily

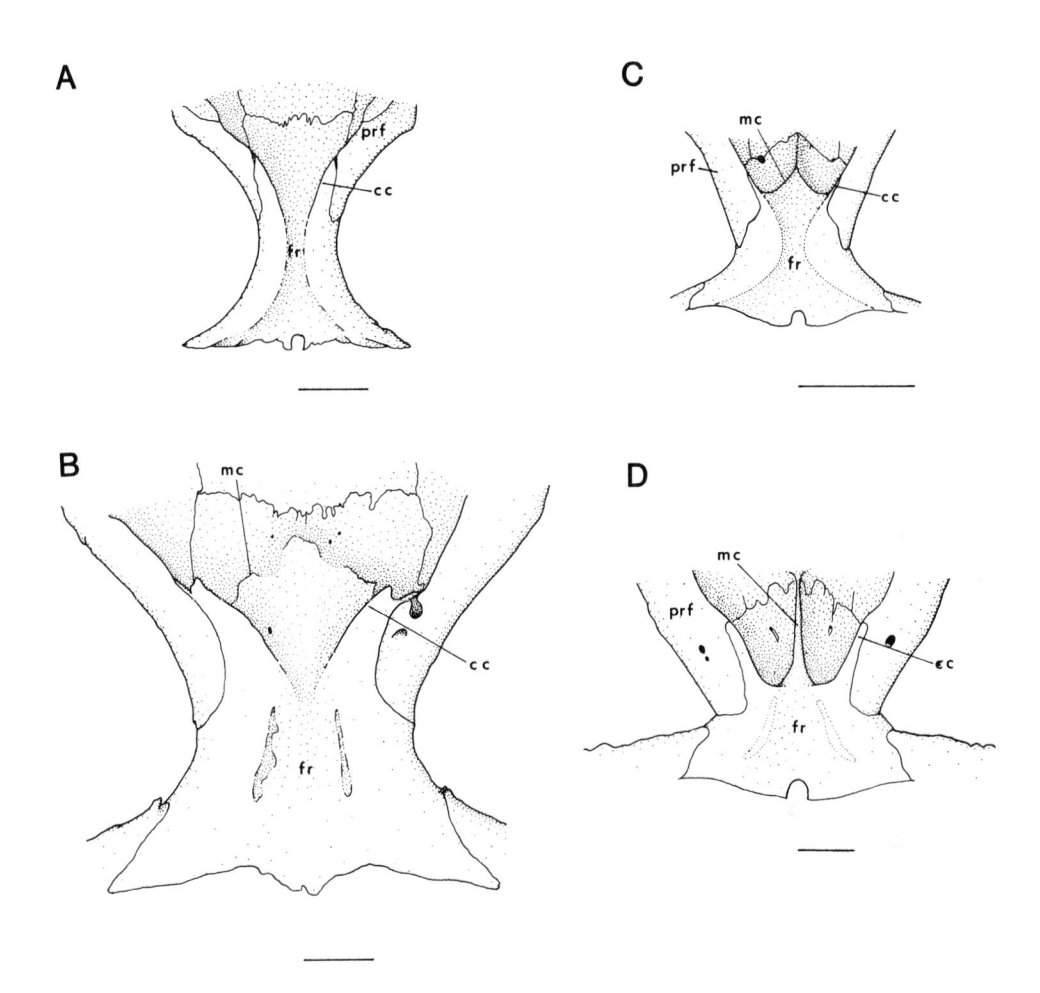

FIG. 12. Ventral views of the frontals of (A) *Ctenosaura pectinata* (RE 641), (B) *Conolophus subcristatus* (AMNH 165756), (C) near-hatching *Amblyrhynchus cristatus* (SDNHM 45157), and (D) adult *A. cristatus* (RE 1387), showing differences in the morphology of the cristae cranii and the development of crests medial to the cristae cranii. Scale equals 0.5 cm. Abbreviations: cc, crista cranii; fr, frontal; mc, medial crest; prf, prefrontal.

distinguished from that of all other iguanines and warrants recognition as a character. All outgroups either lack the medial cristae of the frontals or have them only weakly developed.

The parietal foramen is a small hole that pierces the dermal skull roof above the anterior portion of the brain. It serves as a window in the skull for the parietal eye, a photosensitive organ (Hamasaki, 1968, 1969). The location of the parietal foramen relative to the frontal and parietal bones is variable within iguanines (Table 2). Because the medial portion of the

TABLE 2. Position of the Parietal Foramen

Taxon	N	A	B	C	D
Amblyrhynchus cristatus	19		*100%*		
Brachylophus fasciatus	15		*100*		
vitiensis	4		*100*		
Conolophus pallidus	13		*92*		8%
subcristatus	12		*92*		8
Ctenosaura acanthura	18		*100*		
bakeri	2		*100*		
clarki	9		*67*		33
defensor	1				*100*
hemilopha	18		*67*	11%	22
palearis	2		*50*		*50*
pectinata	14		*86*		14
quinquecarinata	3		*67*		33
similis	18		*100*		
Cyclura carinata	5			*100*	
collei	0	-	-	-	-
cornuta	11		*91*		9
cychlura	7		*100*		
nubila	9		*100*		
pinguis	1		*100*		
ricordii	4		*100*		
rileyi	1		*100*		
Dipsosaurus dorsalis	25	4%	4	12	*80*
Iguana delicatissima	7		*100*		
iguana	23		*100*		
Sauromalus ater	5		*80*		20
australis	2		*50*		*50*
hispidus	26	4	31	23	*42*
obesus	21	5	*48*	24	24
slevini	3		33		*67*
varius	19	16	*32*	*32*	21

Note: Numbers in the last four columns are percentages. Column abbreviations are as follows: N, number of specimens; A, absent; B, within the frontoparietal suture; C, within the frontal but connected to the frontoparietal suture by a suture; D, within the frontal and not connected to the frontoparietal suture. Modes are in italics.

skull roof in the vicinity of the frontoparietal suture ossifies in postembryonic ontogeny, the location of the parietal foramen cannot always be determined in young specimens.

In *Amblyrhynchus, Brachylophus, Conolophus,* and *Iguana*, some species of *Ctenosaura*, and most species of *Cyclura*, this foramen almost always lies within the frontoparietal suture, though it may notch either frontal or parietal more than the other. The parietal foramen is entirely within the frontal in *Cyclura carinata* and *Dipsosaurus*; a suture connecting it with the frontoparietal suture is usually present in the former but usually absent in the latter. The single skull of *Ctenosaura defensor* examined has the parietal foramen located entirely within the frontal. All species of *Sauromalus* and several species of *Ctenosaura* are variable in this character: the parietal foramen in these taxa is commonly found both at the suture between frontal and parietal or entirely within the frontal.

Most members of the outgroups examined in this study have the parietal foramen at the frontoparietal suture, suggesting that this condition is plesiomorphic for iguanines. Exceptions are *Basiliscus* and *Corytophanes*, which have the parietal foramen entirely within the frontal; *Laemanctus*, in which the parietal foramen may be either on the frontoparietal suture or within the frontal; and *Morunasaurus annularis*, in which the parietal foramen appears to be absent. Assuming that fixation of an apomorphic feature is more readily achieved through a polymorphic intermediate condition, I recognized the variable condition as the intermediate stage in a three-state transformation series.

Postfrontals (Figs. 5A, 6A). The postfrontals are small bones confined to the posterodorsal margins of the orbits. The posterior surface of each postfrontal is sutured medially to the frontal and laterally to the postorbital. The postfrontal is invariably present as a discrete element in iguanines, morunasaurs, and *Laemanctus*; it is indistinct (absent or fused) in crotaphytines, oplurines (rarely, a small separate bone is present), and the basiliscines *Corytophanes* and *Basiliscus*.

In iguanines, the lateral portion of the postfrontal may form part of a bony knob along with the postorbital (Figs. 6A), which serves as an attachment point for the skin (Oelrich, 1956). The relative development of this knob varies among iguanine genera. *Amblyrhynchus* and *Brachylophus* have moderate-sized knobs directed mostly laterally. The knob is small or absent in *Conolophus, Ctenosaura, Dipsosaurus,* and *Sauromalus*. *Iguana* and especially *Cyclura* have large, anteriorly directed knobs (Fig. 9A). The relative size of the postfrontal-postorbital knob increases with increasing body size, making it difficult to compare those of animals differing greatly in body size. Because of this problem, I have chosen not to use the variation in the development of this knob as a systematic character.

Postorbitals (Figs. 5A, 6A). The postorbitals of iguanines are paired, triradiate bones situated on the posterolateral sides of the skull just behind the orbits. Their lateral surfaces are often slightly concave. Like those of all iguanians, the postorbitals of iguanines form a major part of the postorbital bar and articulate anteroventrally with the jugals and posteroventrally with the squamosals. Preliminary examination suggested that the relationship of postfrontal to parietal might be a useful systematic character. The postorbital generally ends medially where it contacts the parietal, or it may slightly overlap

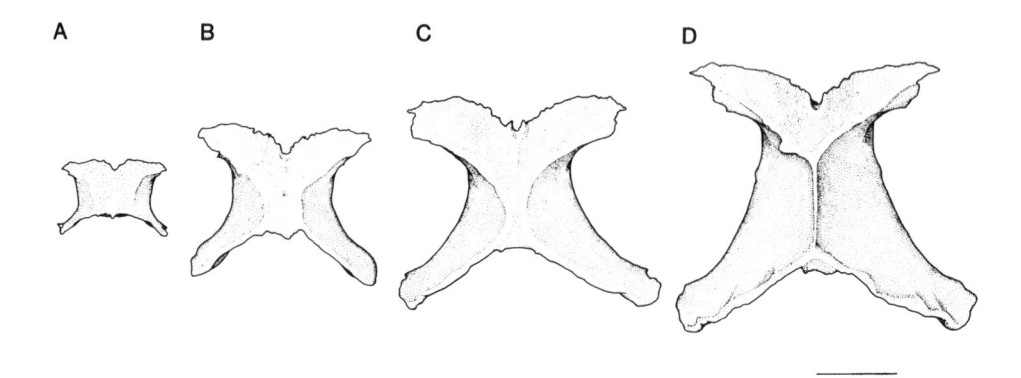

FIG. 13. Dorsal views of the parietals of four *Iguana iguana*-(A) RE 454, condylobasal length = 31.3mm; (B) JMS 713, 59.6mm; (C) RE 424, 71.6mm; (D) RE 489, 80.7mm-showing ontogenetic change in the shape of the parietal roof. Scale equals 1 cm.

the posterior side of the anterolateral process of the parietal. In some iguanines, this overlap is much more extensive, and in others postorbital and parietal fail to contact. Because variation in this feature seems to be greater within the basic taxa than between them, I did not use this variation as a systematic character.

Parietal (Figs. 5A, 6A, 13). The posteriormost bone lying on the midline of the skull roof is the parietal, which is unpaired in postembryonic development. This bone forms a nearly straight transverse suture with the frontal anteriorly, and meets the postorbitals and postfrontals anterolaterally. It has a supratemporal process extending posterolaterally to the complex articulation for the cephalic condyle of the streptostylic quadrate. The adductor muscles of the jaw originate on the dorsolateral surfaces of the parietal. Their area of attachment is set off by distinct crests from the medial and anterior portions of the bone, to which the skin adheres. This latter area, the parietal roof, changes shape during the postembryonic ontogeny of all iguanines. Changes in the shape of the parietal roof are most pronounced in large species (Fig. 13), which undergo the greatest changes in size after hatching. At hatching, the parietal roof is roughly trapezoidal, the lateral crests being widely separated. As the animal grows, the posterior portions of the crests move progressively closer together until they meet, forming a V-shaped roof. Further growth results in elongation of the single median crest formed by the posterior union of the lateral crests, so that the parietal roof takes on the shape of a Y with a growing leg. Different taxa stop at different points along this pathway, and the point of termination is correlated with size. Similar ontogenetic changes in the shape of the parietal roof have been noted in other iguanids (Etheridge, 1959).

The phylogenetic significance of differences in parietal roof shape cannot be adequately assessed without a detailed allometric study, for differences in shape are complicated by differences in the sizes of the organisms being compared. Furthermore, outgroup comparison has limited value here, because no other iguanids get as large as the largest iguanines; and the basiliscines, moderately large iguanids, have highly modified parietals. Nevertheless, a few observations warrant mention. The lateral parietal crests of *Sauromalus* never meet, despite the fact that some of its species (*S. hispidus* and *S. varius*) attain larger sizes than other iguanines in which the crests do eventually meet (e.g., *Brachylophus*). In *Amblyrhynchus*, one of the largest iguanines, the parietal crests do eventually meet, but this occurs at a larger size than in other iguanines, and a Y-shaped roof does not develop. If the complete ontogenetic pathway described above is plesiomorphic for iguanines, then *Sauromalus* and *Amblyrhynchus* exhibit derived conditions. Because of the ambiguities involved in this character, however, I omitted it from the phylogenetic analysis.

A unique feature of the parietal is seen within the genus *Ctenosaura*. In *C. acanthura* and *C. pectinata*, the parietal extends posteriorly, forming a shelf above the braincase (H. M. Smith, 1949, and pers. comm. cited by Ray and Williams, unpublished). This feature suggests that *C. acanthura* and *C. pectinata* form a clade within *Ctenosaura*, and it is used as a systematic character only in an analysis of relationships within that taxon.

Supratemporals (Figs. 5C, 6A). The supratemporals are small bones that form the major part of the articulation for the dorsal end of the streptostylic quadrates. The posterior end of each supratemporal is wedged between four bones: the quadrate ventrally, the squamosal laterodorsally, the parietal dorsally, and the exoccipital medially. In all iguanines, the posterior end of the supratemporal wraps around the ventral edge of the supratemporal process of the parietal. As it extends anteromedially, the supratemporal becomes confined to the posteromedial surface of the supratemporal process. The greater part of the supratemporal lies on this posteromedial surface. In most other lizards, the supratemporal lies primarily on the anterolateral surface of the supratemporal process of the parietal for its entire length. Sometimes it extends along the ventral edge of the supratemporal process and bears medial and lateral portions of approximately equal size. As far as I am aware, oplurines, *Enyalius*, and mosasaurs are the only other lizards in which the greater portion of the supratemporal lies on the posteromedial surface of the supratemporal process of the parietal. Therefore, I interpret this condition as an iguanine synapomorphy.

The anterior extent of the supratemporal varies within Iguaninae. In most, the supratemporal extends forward at least halfway across the posterior temporal fossa. This condition occurs also in basiliscines, crotaphytines, morunasaurs, and oplurines, and is therefore taken to be plesiomorphic for iguanines. The supratemporal of *Conolophus* has apparently been reduced phylogenetically; it sometimes reaches halfway across the posterior temporal fossa, but generally falls short of this point.

Maxillae (Figs. 5A,B, 6A, 14). The maxillae are paired bones that bear most of the upper marginal teeth. They are roughly triangular and lie on the anterior sides of the skull, where they meet the premaxilla anteriorly, the nasals and prefrontals dorsally, the lacrimals and jugals posterodorsally, and the ectopterygoid posteriorly. A number of supralabial foramina pierce the maxillae in a row parallel to its ventral border. Compared to the

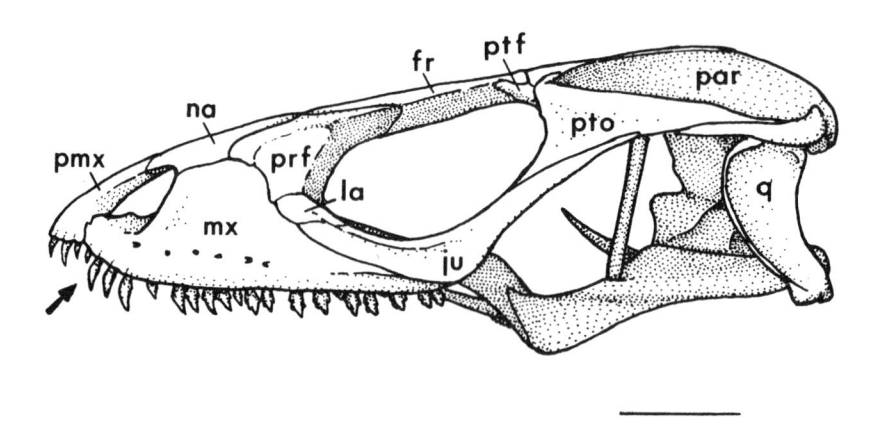

FIG. 14. Lateral view of the skull of *Ctenosaura similis* (MCZ 21742), showing the dorsal curvature of the premaxillary process of the maxilla (arrow). Scale equals 1 cm. Abbreviations: fr, frontal; ju, jugal; la, lacrimal; mx, maxilla; na, nasal; par, parietal; pmx, premaxilla; prf, prefrontal; ptf, postfrontal; pto, postorbital; q, quadrate.

supralabial foramina of other iguanines, those of *Amblyrhynchus* seem to lie slightly higher on the maxillae above a rounded ridge that is not seen in any other iguanine or in any outgroup examined in this study. There is also variation in the relative size of the supralabial foramina within Iguaninae, with large ones being found in some species of *Cyclura*, especially *C. cychlura*. This variation is not useful for examining relationships among the basic taxa used in this study, since it occurs in only some *Cyclura*.

The maxillae of *Ctenosaura* are unique in that the premaxillary processes in this genus curve dorsally, so that the premaxilla and the anterior portions of the maxillae are higher than the rest of the upper jaw margin, giving the skull a sneering appearance (Fig. 14). The teeth in this region form large, curved fangs. The dorsal displacement of the anterior end of the tooth row increases allometrically both within and between species of *Ctenosaura*, being less pronounced in juveniles of large species and in adults and juveniles of small species. In all other iguanines, the entire upper jaw margin lies in a single horizontal plane or is only slightly elevated anteriorly (Fig. 6A). Although no other iguanines nor any of the outgroups exhibit as pronounced a curvature of the premaxillary process of the maxilla as that seen in large *Ctenosaura*, many taxa are difficult to compare because of their small size and the allometric change seen in this feature.

Lacrimals (Figs. 5A, 6A, 14). The lacrimals of iguanines are small bones situated at the anterior corner of each orbit. In *Amblyrhynchus* these bones are relatively small compared to those of other iguanines. *Conolophus* also has relatively small lacrimals, intermediate in size between those of *Amblyrhynchus* and the smallest ones seen in other iguanines. All other iguanines have relatively large lacrimals whose size and shape vary among the genera. This variation ranges from the long, curved bones of *Brachylophus*

(Fig. 6A) and *Dipsosaurus* to the almost square ones of *Ctenosaura* (Fig. 14), *Cyclura*, and *Iguana* (those of *Sauromalus* are intermediate). Although the extreme lacrimal morphologies in iguanines other than *Amblyrhynchus* and *Conolophus* are very different, the variation between them is more or less continuous.

Basiliscines, crotaphytines, and oplurines all have relatively large lacrimals, but those of morunasaurs are relatively small and may even be absent in some *Morunasaurus*. Thus, it appears that a small lacrimal is apomorphic within iguanines, although the evidence is not completely unambiguous. If so, then the small lacrimals of morunasaurs must be convergent.

Jugals (Figs. 5A, 6A, 14). The iguanine jugals form the ventral margins of the orbits and are sutured anteriorly with the lacrimals, anteroventrally with the maxillae, medially with the ectopterygoids, and posterodorsally with the postorbitals. Each jugal extends posteriorly along the ventral border of the postorbital and variably contacts the squamosal on the ventral surface of the upper temporal arch. This contact appears to be too variable within genera to serve as a character for examining their interrelationships.

Squamosals (Figs. 5A,C, 6A, 15). At the posterior end of each temporal arch, behind the postorbitals, lie the squamosals. The shape of the squamosal is variable, ranging from long and thin in *Ctenosaura* and *Sauromalus* to short and wide in *Amblyrhynchus*; the remaining genera are intermediate. At its posterior end, the squamosal bears two processes: a dorsal process that meets the supratemporal process of the parietal, and a ventral process or peg directed towards the quadrate. The relative size of the ventral process is variable, being more strongly developed in *Amblyrhynchus* and *Iguana* than in the other genera (Fig. 15). In these two genera, however, the relationship of the ventral process of the squamosal to the quadrate is different. As in most iguanines (Figs. 6A, 15A), the ventral process of *Amblyrhynchus* (Fig. 15B) lies against the anterior edge of the cephalic condyle of the quadrate, projecting into a gap or hole between this condyle and the dorsal portion of the tympanic crest of the quadrate. In *Iguana*, the ventral process abuts directly against the top of the tympanic crest (Fig. 15C), presumably reducing the mobility of the quadrate. *Cyclura* possesses a potentially incipient stage to the condition seen in *Iguana*. Its squamosal also abuts against the tympanic crest of the quadrate, but does so more weakly because the ventral process is not as large (Fig. 15D).

The relationship of the ventral process of the squamosal to the quadrate, and the relative size of this process in basiliscines, crotaphytines, morunasaurs, and oplurines, are very similar to those of *Amblyrhynchus*, suggesting that this condition is plesiomorphic for iguanines. The unique articulation between squamosal and quadrate in *Cyclura* and *Iguana* suggests that the large ventral process in *Iguana* may not be homologous with those of *Amblyrhynchus* and non-iguanines. In order to maintain objectivity, however, I did not assume that such was the case.

Quadrates (Figs. 5B,C, 6A, 15). The quadrates lie at the posteroventral corners of the skull. They are streptostylic and are important in jaw mechanics and feeding (Rieppel, 1978; K. K. Smith, 1980). Ventrally, each quadrate forms the articulation of the skull with the articular bone of the mandible, and it also articulates dorsally with the squamosal,

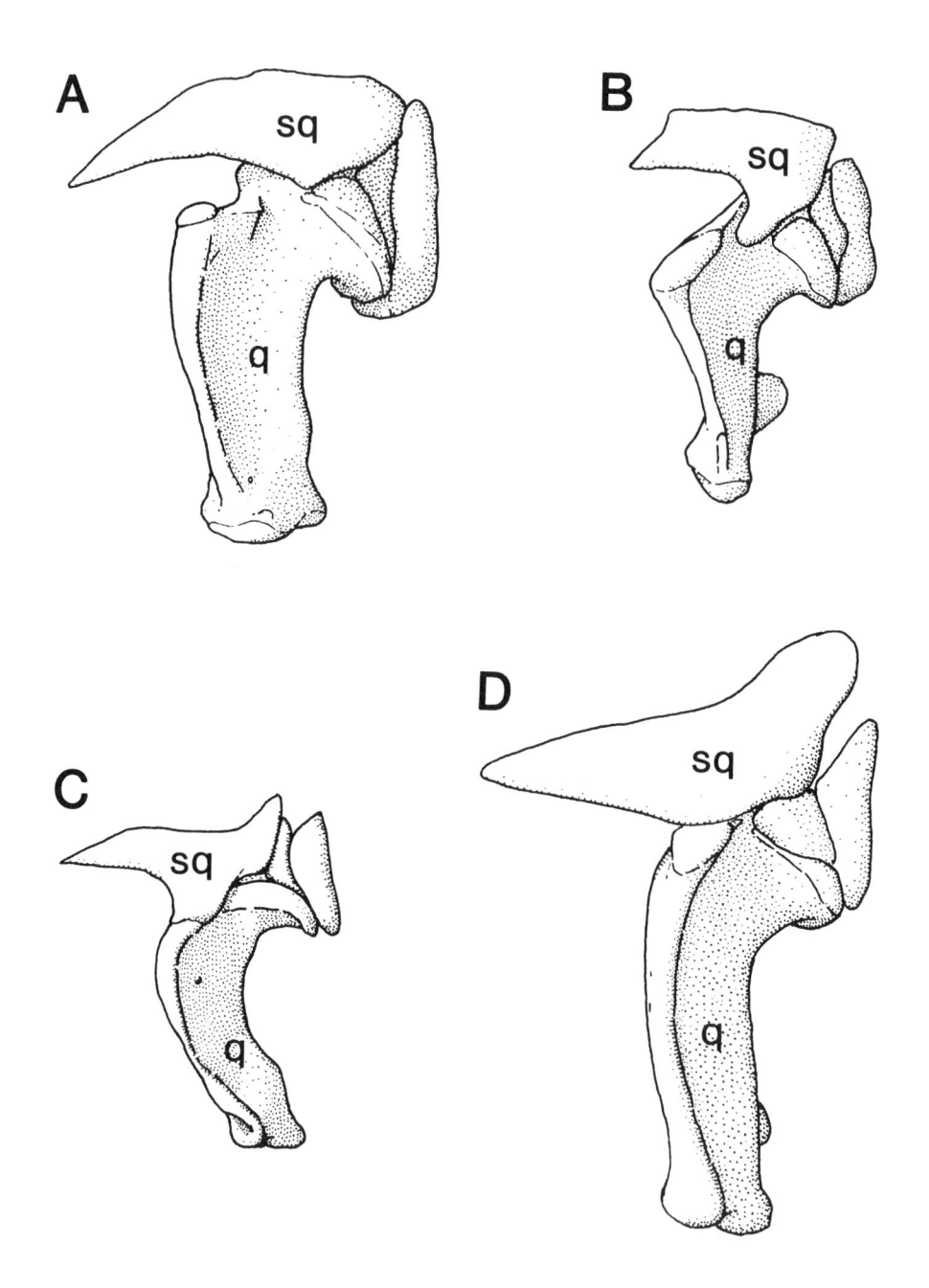

FIG. 15. Lateral views of the posterolateral corners of the skulls of (A) *Conolophus pallidus* (RE 439), (B) *Amblyrhynchus cristatus* (RE 1387), (C) *Iguana iguana* (RE 1006), and (D) *Cyclura cornuta* (RE 383), showing differences in the size of the ventral process of the squamosal and the articulation between squamosal (sq) and quadrate (q).

dorsomedially with the supratemporal, and ventromedially with the pterygoid. The quadrate is bowed forward, and its lateral edge, the tympanic crest, supports the anterior edge of the external tympanic membrane. In small species and juveniles of larger species, the quadrates tilt backwards with their articular condyles lying anterior to the transverse plane at the posterior end of the occipital condyle. During ontogeny the articular condyles move posteriorly, eventually coming to lie posterior to the occipital condyle and giving the quadrates a forward tilt. A related change occurs in the pterygoids, which must elongate posteriorly as the articular condyles move backwards if they are to remain connected to the quadrates. Variation in the relationship of the squamosal to the quadrate has been described above.

PALATE

The iguanine palate (Fig. 5B) consists of four pairs of dermal bones that form the floor of the skull proper and the roof of the mouth: vomers, palatines, pterygoids, and ectopterygoids.

Vomers (Fig. 5B). The vomers are the anteriormost bones of the palatal complex. They are paired elements lying on either side of the midline and articulating with the maxillae and premaxilla anteriorly, the septomaxillae dorsally, and the palatines posteriorly. The shape of the vomers differs both among iguanine genera and among iguanines and various outgroups; however, I was unable to partition this variation into character states and to assess its polarity.

Palatines (Figs. 5B, 16, 17, 18). Just posterior to the vomers lie the paired palatines, which form the major portion of the palate. These bones also make up the anterior floor of the orbits and the posterior floor of the nasal capsules. Each palatine has three processes: the vomerine process anteriorly, the maxillary process laterally, and the pterygoid process posteriorly.

In *Brachylophus, Ctenosaura, Cyclura, Dipsosaurus, Iguana,* and *Sauromalus,* as well as in basiliscines, crotaphytines, oplurines and morunasaurs, the vomerine process of the palatine bears a low ridge that extends longitudinally along its dorsomedial edge (Fig. 16A). This ridge bends laterally at the posterior end of the nasal capsule. In place of the low ridge, *Amblyrhynchus* and *Conolophus* have a high crest and thus greater bony separation of the nasal capsules (Fig. 16B).

Behind the maxillary process, the palatine forms the medial border of the suborbital fenestra (inferior orbital foramen of Oelrich, 1956). In this region, the palatines of some *Sauromalus* differ from those of all other iguanines and all outgroups examined in this study in that their lateral borders are fringed along the anterior margins of the suborbital fenestrae. I have observed this feature in several species of *Sauromalus,* but its presence always appears to be variable.

The infraorbital foramen, which transmits the superior alveolar nerve and artery (Oelrich, 1956), pierces the anterior orbital wall in the region of the maxillary process of the palatine. The exact position of the foramen relative to this process varies within

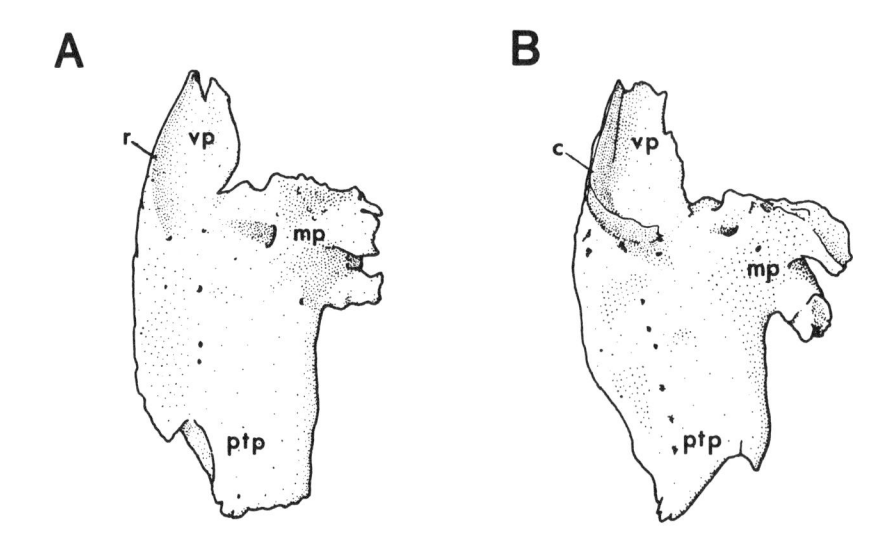

FIG. 16. Posterodorsal views of disarticulated right palatines of (A) *Iguana delicatissima* (MCZ 75388) and (B) *Conolophus subcristatus* (AMNH 110168), contrasting the low dorsomedial ridge in the former with the high dorsomedial crest in the latter. Abbreviations: c, dorsomedial crest; mp, maxillary process; ptp, pterygoid process; r, dorsomedial ridge; vp, vomerine process.

iguanines, and five categories can be recognized (Fig. 17): (1) entirely within the palatine without a suture connecting the foramen to the lateral edge of the palatine (Fig. 17A); (2) within the palatine with a suture running from the foramen to the lateral edge of the palatine (Fig. 17B) (individuals falling within this category vary considerably in the distance from the foramen to the lateral edge of the palatine and thus the length of the suture); (3) between the palatine and the jugal (Fig. 17C) (sometimes the lacrimal also contributes to the border of the foramen); (4) between palatine and maxilla with the portion of the palatine directly posterior to the foramen large but not extending laterally to contact the jugal (Fig. 17D); (5) between palatine and maxilla, with the portion of the palatine directly posterior to the foramen small or absent (Fig. 17E).

Intrageneric and intraspecific variation in the position of the iguanine infraorbital foramen is great: many genera and species exhibit more than one of the five conditions described above. Nevertheless, sufficient differences exist among genera that some of the variation can be used in phylogenetic analysis. Most iguanines (*Amblyrhynchus*, *Conolophus*, *Ctenosaura*, *Cyclura*, *Iguana*, and *Sauromalus*) commonly exhibit condition 3, in which the infraorbital foramen is located at the suture between palatine and jugal. *Conolophus*, *Cyclura*, and *Iguana* are relatively constant in the position of this foramen: the great majority of individuals in these three genera exhibit condition 3. *Amblyrhynchus*, *Ctenosaura*, and *Sauromalus* are more variable. Though condition 3 occurs commonly in all three genera, in *Amblyrhynchus* and *Ctenosaura* the infraorbital foramen of many individuals is entirely within the palatine. A suture connecting the foramen to the lateral

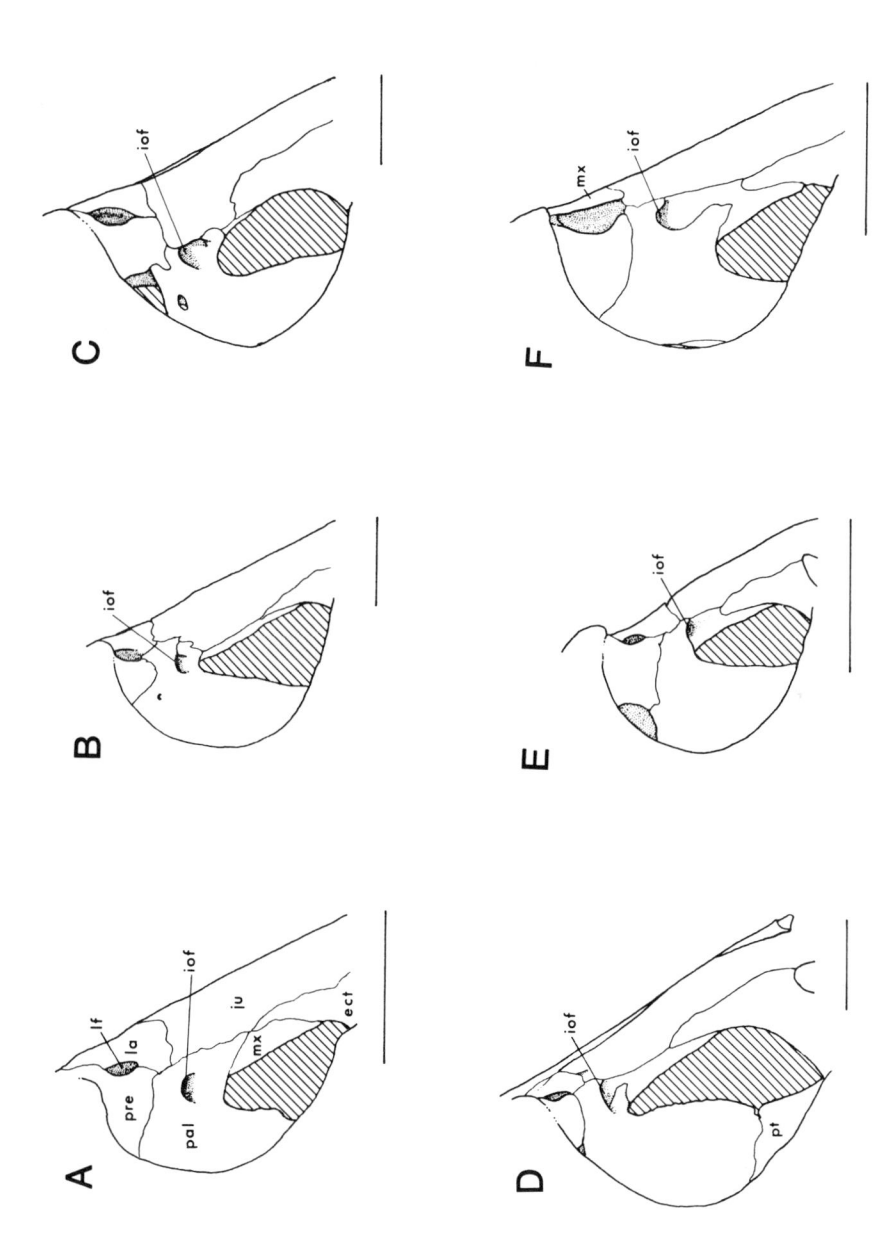

FIG. 17. Posterodorsal views of the right orbits of (A) *Brachylophus fasciatus* (RE 1888), (B) *Ctenosaura similis* (MCZ 36830), (C) *Conolophus pallidus* (MCZ 79772), (D) *Sauromalus varius* (RE 308), (E) *Dipsosaurus dorsalis* (RE 356), and (F) *Morunasaurus annularis* (RE 1956), showing differences in the position of the infraorbital foramen. Scale equals 0.5 cm. Abbreviations: ect, ectopterygoid; iof, infraorbital foramen; ju, jugal; la, lacrimal; lf, lacrimal foramen; mx, maxilla; pal, palatine; pre, prefrontal; pt, pterygoid.

edge of the palatine is generally present (condition 2). *Sauromalus obesus* is similar to *Amblyrhynchus* and *Ctenosaura* in this regard, but specimens of *S. ater*, *S. hispidus*, and *S. varius* exhibit condition 4, in which the maxilla contributes to the ventral rim of the foramen, rather than condition 2 (samples of other species of *Sauromalus* are too small upon which to base generalizations).

Brachylophus and *Dipsosaurus* are unique among iguanines in the positions of their infraorbital foramina, though at different ends of the morphological spectrum. In *Brachylophus*, the infraorbital foramen is entirely within the palatine. A suture extending from the foramen to the lateral edge of the maxillary process of the palatine (condition 2) was observed in all four *B. vitiensis* examined but was absent (condition 1) in over half of the specimens of *B. fasciatus*. *Dipsosaurus* is the only iguanine that commonly exhibits condition 5, in which the infraorbital foramen emerges between palatine and maxilla. In some specimens, a small posteriorly or laterally directed process is present at the medial edge of the foramen; in others it is absent. When present, the process is smaller than that seen in other iguanines (some *Sauromalus*) in which this process fails to contact the jugal laterally.

Because of the high intrageneric variation in the position of the infraorbital foramen, I recognized three characters each with one apomorphic state rather than one character with four or five: one for the size of the portion of the palatine immediately posterior (or posteromedial) to the infraorbital foramen, a second for the presence or absence of contact between this part of the palatine and the jugal, and a third for whether or not the infraorbital foramen lies entirely within the palatine.

The infraorbital foramina of the four outgroups examined in this study generally differ from any of those seen in iguanines. Basiliscines and morunasaurs exhibit a condition similar to that described above as condition 4, but the process of the palatine at the medial edge of the infraorbital foramen is directed posteriorly rather than laterally (Fig. 17F). *Chalarodon* and some *Oplurus* possess condition 5, while other *Oplurus* possess the condition described for basiliscines and morunasaurs. Individual crotaphytines may also exhibit the basiliscine-morunasaur condition, but in other individuals the infraorbital foramen is located between palatine and jugal as in some iguanines. In the latter case, however, the contact of the posteriorly directed process of the palatine with the jugal results from extensive medial development of the jugal, rather than from lateral extension of the process of the palatine as in iguanines.

The differences between iguanines and the four outgroups indicate either that some morphological change occurred between the most recent common ancestor of iguanines and their closest relatives among these four outgroups or that no living iguanine species is characterized by the condition that was present in the most recent common ancestor of the group (though some individual specimens may be). Nevertheless, differences between iguanines and the outgroups are minor enough that the polarities of all three characters can be assessed. Because no iguanines possess the same morphology of the infraorbital foramen seen in the outgroups, no iguanine is scored plesiomorphic for all three characters.

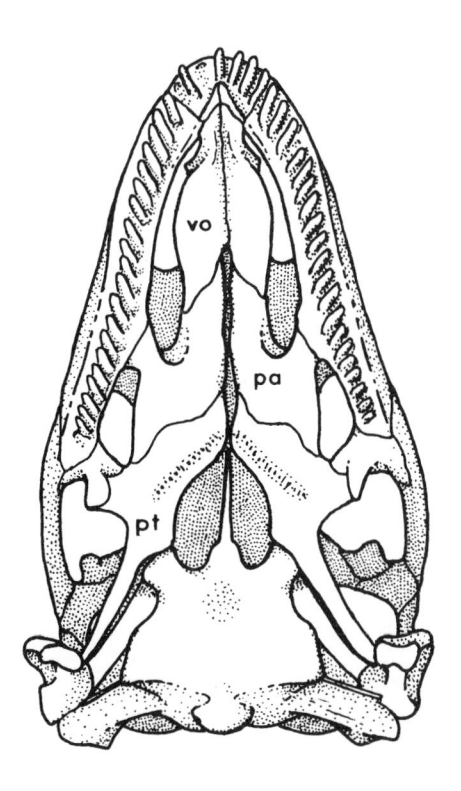

FIG. 18. Ventral view of the skull of *Iguana delicatissima* (MCZ 16157), showing the medial curvature of the pterygoids and concomitant abrupt narrowing of the pyriform recess. Scale equals 1 cm. Abbreviations: pa, palatine; pt, pterygoid; vo, vomer.

Pterygoids (Figs. 5B,C, 6A, 18). These paired bones are the posteriormost palatal elements. Each pterygoid bears three processes: an anteriorly directed palatine process, an anterolaterally directed transverse process, and a longitudinally compressed and posteriorly directed quadrate process. The ventral surface of the palatine process often bears small teeth. Anterior to the pterygoid notches, where the basipterygoid processes of the basisphenoid articulate with the pterygoids, the medial edges of the pterygoids of most iguanines curve towards the midline, resulting in a sudden narrowing of the pyriform recess (interpterygoid vacuity) (Fig. 18). In contrast, the medial edge of the pterygoids in *Brachylophus* is relatively straight, and the pyriform recess narrows more gradually from posterior to anterior (Fig. 5B).

Outgroup comparison suggests that the condition seen in *Brachylophus* is plesiomorphic. Among the four outgroups, only crotaphytines exhibit the strongly curved medial borders of the pterygoids, though a moderate curvature occurs in some oplurines. Thus, depending upon the relationships among ingroup and outgroups, either the polarity of this character will be equivocal, or the interpretation that the relatively straight medial border of the pterygoids is plesiomorphic will be favored.

Ectopterygoids (Figs. 5A,B, 6A). Each ectopterygoid lies at the posterior margin of the suborbital fenestra forming a brace between the jugal and maxilla anterolaterally and the pterygoid posteromedially. Near the posteromedial corner of the suborbital fenestra, the ectopterygoid may contact the palatine, usually on the dorsal surface of the palatal bones. Contact between ectopterygoid and palatine in this region is the common condition only in *Conolophus* among iguanines, and occurs in about half of the *Iguana delicatissima* examined. This contact occurs rarely in some other iguanine species. Ectopterygoid-palatine contact in this region was not observed in any of the four outgroups and is therefore considered apomorphic.

The ectopterygoid may also contact the palatine near the anterolateral corner of the suborbital fenestra. This condition is clearly derived for iguanines on the basis of outgroup comparison, but does not appear to be characteristic of any iguanine species. Only *Amblyrhynchus* exhibits the anterolateral ectopterygoid-palatine contact commonly, but even here it occurs in less than half of the specimens examined. Because the apomorphic state of this character is not characteristic of any iguanine species and because diagnostic apomorphies of *Amblyrhynchus* are plentiful, I have chosen to ignore this character in the phylogenetic analysis.

BRAINCASE

The iguanine braincase (Figs. 5A,B, 6A), or neurocranium, is composed of four pairs of endochondral bones-orbitosphenoids, prootics, opisthotics, and exoccipitals-and three unpaired ones-basisphenoid, basioccipital, and supraoccipital. The parasphenoid, a dermal bone, is also described here because of its intimate association with the basisphenoid. Parasphenoid and basisphenoid as well as exoccipitals and opisthotics are fused to each other even in juveniles, and all other elements except orbitosphenoids fuse with neighboring braincase elements late in ontogeny. In some very large specimens, even the orbitosphenoids are fused with one another. Although the stapes and epipterygoids are splanchnocranial elements, they are included in this section because of their close associations with the braincase.

Orbitosphenoids (Fig. 19). The orbitosphenoids are paired, crescent-shaped bones lying within the membranes that separate the brain cavity from the orbits. Each orbitosphenoid is continuous with five orbital cartilages: the septal cartilage and planum supraseptale anterodorsally, the pila accessoria and pila antotica posterodorsally, and the hypochiasmatic cartilage ventrally (Oelrich, 1956). Although consistent differences in the shapes of the orbitosphenoids exist between iguanine taxa, these differences seem to be related to differences in body size. In large iguanines, the orbitosphenoids undergo considerable ontogenetic changes in shape resulting from progressive outward ossification

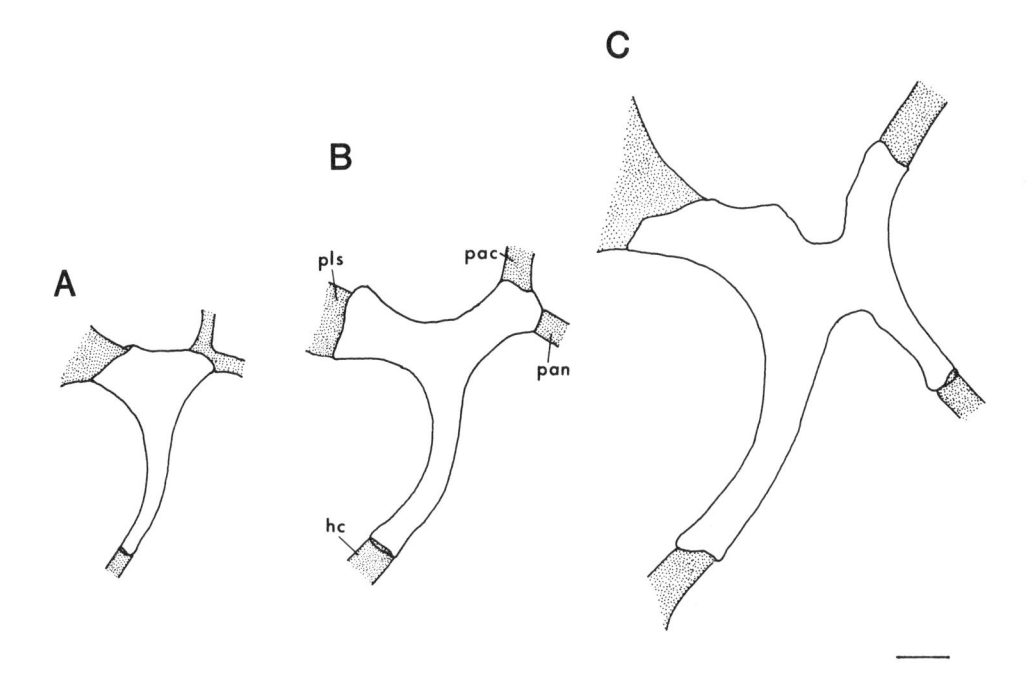

FIG. 19. Anterolateral views of the left orbitosphenoids of three *Iguana iguana*-(A) RE 454, (B) JMS 245, (C) JMS 713-showing ontogenetic change in the shape of these bones resulting from progressive ossification outward along the orbital cartilages. Scale equals 1 mm. Abbreviations: hc, hypochiasmatic cartilage; pac, pila accessoria; pan, pila antotica; pls, planum supraseptale.

along the orbital cartilages (Fig. 19). Thus, the posterodorsal edge of each orbitosphenoid first develops a posterior process where it joins the pila accessoria and pila antotica, and this process later bifurcates following the two diverging orbital cartilages. The ventral and anterodorsal ends of the bone elongate by a similar process and, in the case of the latter, the two orbitosphenoids may eventually meet and fuse at the midline. Small iguanines generally fail to develop the bifurcating posterodorsal processes of the orbitosphenoids seen in adults of larger species, and I have never observed medial fusion of the two bones at their anterodorsal ends in small iguanines.

Epipterygoids (Fig. 6A). The epipterygoids are thin, rod-shaped bones extending from the palate to the skull roof. Ventrally, the epipterygoids sit in depressions in the dorsal surfaces of the palatines, but dorsally their articulations with the parietal are either weak or lacking. I found no differences in epipterygoid morphology among iguanine genera that might serve as systematic characters.

Prootics (Fig. 6A). The paired prootics form the lateral walls of the neurocranium. They are sutured to the supraoccipital dorsomedially, to the exoccipitals posteriorly, to the basioccipital posteroventrally, and to the basisphenoid ventromedially. Although the

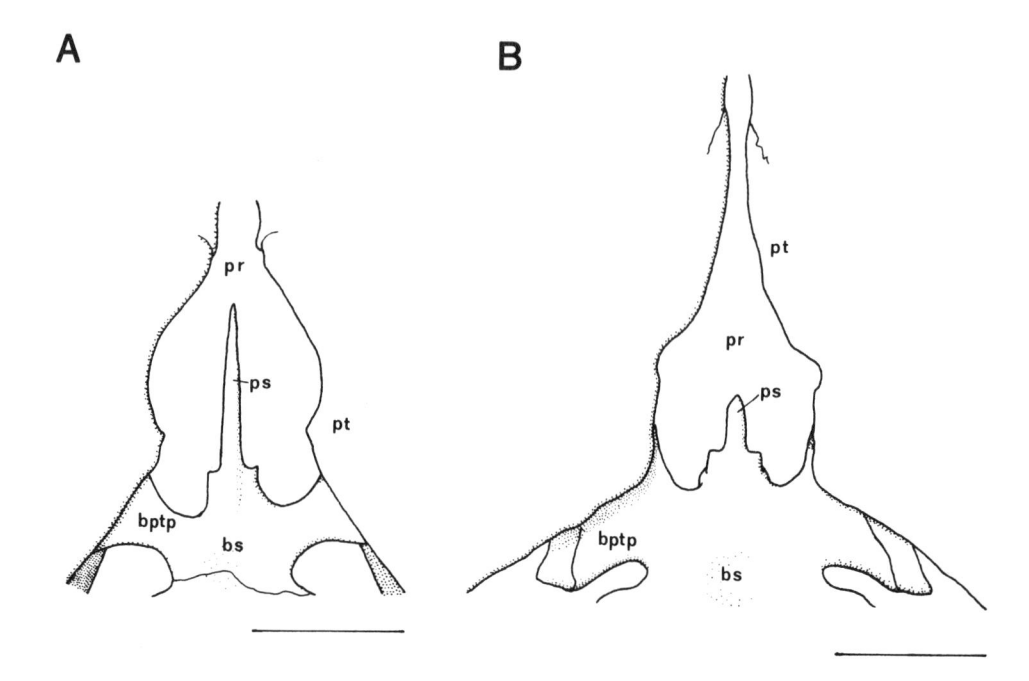

FIG. 20. Ventral views of the posterior portion of the palate and anterior portion of the braincase of (A) *Sauromalus varius* (RE 308) and (B) *Amblyrhynchus cristatus* (RE 1508), showing differences in the length of the parasphenoid rostrum. Scale equals 1 cm. Abbreviations: bptp, basipterygoid process; bs, parabasisphenoid; pr, pyriform recess; ps, parasphenoid rostrum; pt, pterygoid.

morphology of the prootics is complex, I have found no characters in these bones that might serve to elucidate relationships among the basic taxa used in this study.

Parabasisphenoid (Figs. 5B, 6A, 20, 21). Because the parasphenoid and basisphenoid of iguanines are always fused postembryonically, I describe them as a single element. The parasphenoid rostrum extends anteriorly like a thin, flat blade from the main body of the parabasisphenoid on the midline. Compared to those of all other iguanines as well as those of basiliscines, crotaphytines, morunasaurs, and oplurines, the parasphenoid rostrum of *Amblyrhynchus* is relatively short (Fig. 20). Even the parasphenoid rostra of other short-skulled taxa, such as the basiliscine *Corytophanes*, are much longer.

The main body of the parabasisphenoid is an unpaired median bone that forms the anterior floor of the brain cavity. It is sutured with the prootics laterally and with the basioccipital posteriorly. Anterolaterally, two large basipterygoid processes meet the anteromedial surfaces of the quadrate processes of the pterygoids at the pterygoid notches, forming a movable joint between palate and braincase.

Boulenger (1890) first noted variation in the form of the parabasisphenoid (Fig. 21) among different iguanines. In most iguanines, the ventrolateral edges of the parabasisphenoid, the cristae ventrolaterales, are strongly constricted behind the

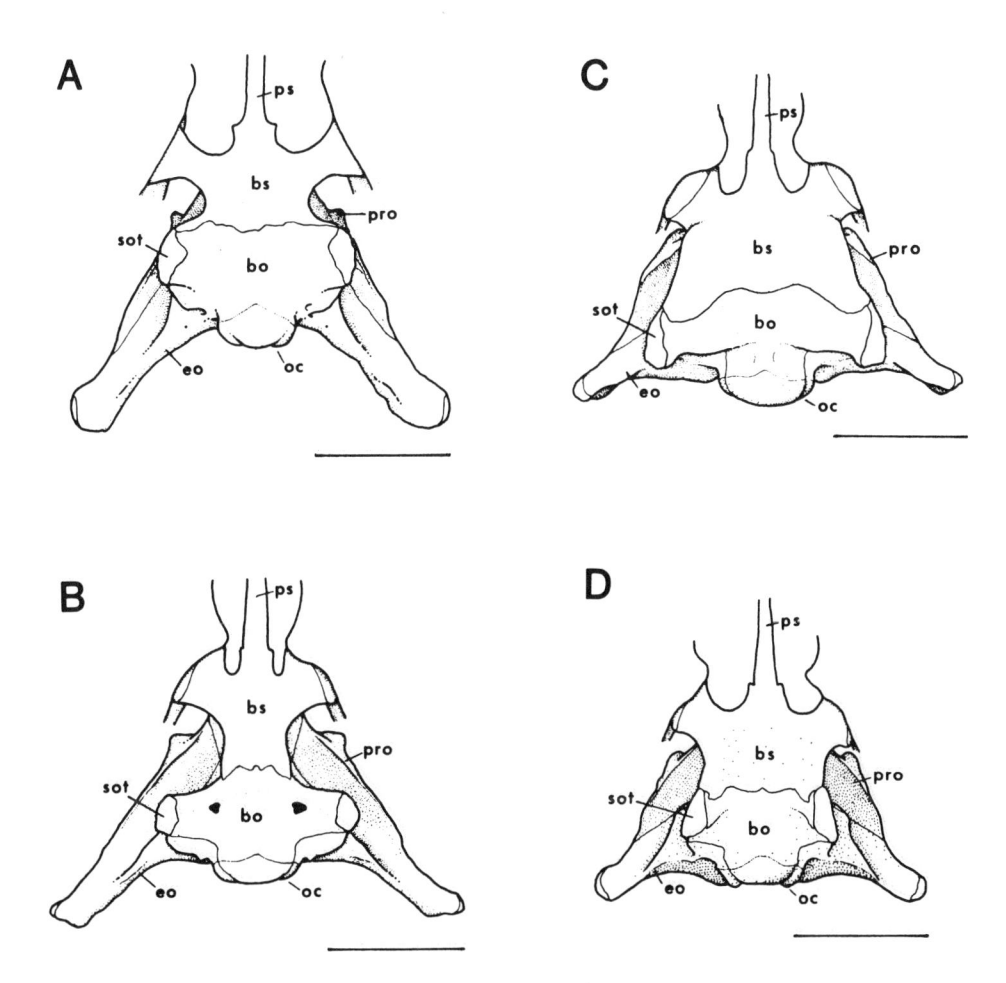

FIG. 21. Ventral views of the neurocrania of (A) *Sauromalus varius* (RE 451), (B) *Ctenosaura hemilopha* (RE 325), (C) *Iguana iguana* (RE 1006), and (D) *Cyclura nubila* (RE 337), showing differences in the width of the parabasisphenoid and the size of its posterolateral processes. Scale equals 1 cm. Abbreviations: bo, basioccipital; bs, parabasisphenoid; eo, exoccipital-opisthotic; oc, occipital condyle; pro, prootic; ps, parasphenoid rostrum; sot, spheno-occipital tubercle.

basipterygoid processes, giving the ventral outline of the braincase roughly the shape of an hourglass (Fig. 21A,B). In contrast, the cristae ventrolaterales of *Iguana* are widely separated, extending in almost straight lines from the basipterygoid processes posteriorly to the spheno-occipital tubercles and giving the ventral outline of the braincase the shape of a box (Fig. 21C). *Cyclura* is variable in this character, though all species have relatively broad parabasisphenoids (Fig. 21D) compared to those of most other iguanines. *C. carinata* has the narrowest basisphenoid, while that of *C. pinguis* is at least as wide as that of some *Iguana delicatissima*; other species are intermediate. In at least some of those

Cyclura with wide parabasisphenoids, this bone becomes relatively wider during postembryonic ontogeny. All basiliscines, crotaphytines, morunasaurs, and oplurines have the parabasisphenoid strongly constricted behind the basipterygoid processes, indicating that this condition is plesiomorphic for iguanines.

A second part of the iguanine parabasisphenoid exhibits two distinct morphologies that are constant within genera. The parabasisphenoids of all iguanines except *Ctenosaura* bear large posterolateral processes that extend along the anterolateral edges of the lateral processes of the basioccipital, reaching or closely approaching the spheno-occipital tubercles (Fig. 21A,C,D). In *Cyclura* (Fig. 21D) and especially in *Iguana* (Fig. 21C), widening of the parabasisphenoid obliterates the distinctness of its posterolateral processes; their existence is inferred from the lateral extent of the parabasisphenoid along the lateral processes of the basioccipital. Unlike other iguanines, the posterolateral processes of the parabasisphenoid are very short or absent in *Ctenosaura* (Fig. 21B), a condition that may be related to the elongation of the skull in this taxon. Only *Crotaphytus* (but not *Gambelia*) among the outgroups examined exhibits a condition similar to that of *Ctenosaura*; therefore, I considered the possession of long posterolateral processes of the parabasisphenoid to be plesiomorphic.

Basioccipital (Figs. 5B, 21). The basioccipital forms the posterior floor of the brain cavity and makes up the large medial portion of the occipital condyle. It bears prominent ventrolaterally directed lateral processes that are capped by the spheno-occipital tubercles. These tubercles fuse to the lateral processes late in ontogeny. The basioccipital is sutured to the exoccipitals dorsolaterally, to the prootics anterolaterally, and to the parabasisphenoid anteriorly. Although iguanine basioccipital morphology is variable, I found no obvious characters that bear on intergeneric relationships.

Exoccipitals and Opisthotics (Figs. 5C, 21). The exoccipitals are indistinguishably fused to the opisthotics in postembryonic developmental stages of all iguanines. These compound bones form the posterior sides of the brain cavity and the lateral edges of the foramen magnum. They meet the supraoccipital dorsomedially, the prootics anteriorly, and the basioccipital ventromedially. The paroccipital processes of the opisthotics extend laterally to contact the supratemporals, bracing the posterolateral corners of the skull. The relative length of the paraoccipital processes varies among iguanine genera, but differences are complicated by positive allometry of this feature both within and among taxa (though the correlation is less precise in the latter case). Apparently the braincase widens more slowly than the skull as a whole. As the paraoccipital processes elongate, they also become more posteriorly oriented.

Each exoccipital-opisthotic bears two prominent crests laterally: the crista interfenestralis, which lies between the fenestra ovalis and the fenestra rotunda; and the crista tuberalis, which bounds the antrum of the fenestra rotunda posteriorly. Variation exists in the degree to which the crista tuberalis slants inward dorsally and to which it obscures the crista interfenestralis in posterior view, but this variation is too great within iguanine genera to be useful for inferring relationships among them. *Dipsosaurus* is unique among iguanines in possessing a sharp, laterally directed point on each crista interfenestralis. Although this process is absent or very small in basiliscines, crotaphytines, morunasaurs, and oplurines, it is present in some sceloporines. I consider *Dipsosaurus* and these sceloporines to have developed a pointed process on the crista interfenestralis convergently.

Stapes. The stapes, or columella, is a sound-transmitting bone that extends from the fenestra ovalis (foramen ovale of Oelrich, 1956) in the braincase to a point just behind the posterior crest of the quadrate. In life it is attached to the external tympanic membrane via a cartilaginous extracolumella, which is often damaged during skeletal preparation. The stapes of *Amblyrhynchus* is robust compared to those of all other iguanines and most other iguanids, although some sceloporines also have a thick stapes (Axtell, 1958; Earle, 1962).

MANDIBLE

There are seven bones present in the mandibles of all iguanines (Fig. 6B,C); from anterior to posterior these are: dentary, splenial, coronoid, angular, surangular, prearticular, and articular. The articular is a splanchnocranial endochondral bone; the remaining bones are dermal. In some noniguanine iguanids, either splenial or both splenial and angular may be absent (Etheridge and de Queiroz, 1988).

Dentary (Figs. 6B,C, 22). The dentary is the anteriormost bone in the mandible and extends posteriorly to about the level of the apex of the coronoid. It is the only tooth-bearing bone in the lower jaw. Anterior to the splenial, Meckel's cartilage, which extends from the articular bone to the anterior end of the mandible, is completely enclosed in a bony tube formed by the dentary. In some other iguanids (e.g., morunasaurs) the groove for Meckel's cartilage is completely open lingually, while in others (e.g., crotaphytines) the dorsal and ventral edges of the groove meet to close the tube but remain separated by a suture. In one late embryo of *Amblyrhynchus* (SDNHM 45156), Meckel's groove is closed but retains a suture; however, in all postembryonic iguanines the upper and lower dentary portions of Meckel's groove are closed and fused.

A series of mental foramina are positioned along the labial face of the anterior half of the dentary. In all iguanines except *Amblyrhynchus* and in all outgroups examined, these foramina lie in a line about halfway between the dorsal and ventral edges of the dentary, and the dorsal edge of the dentary where it meets the coronoid is approximately level with the dorsal border of the surangular just posterior to the coronoid (Fig. 22A). The dorsal border of the dentary in *Amblyrhynchus* is high, well above the level of the dorsal border of the surangular, and the row of mental foramina lies more than halfway down the labial surface of the dentary (Fig. 22B).

Splenial (Fig. 6B,C, 23). The exposed portion of the splenial is roughly diamond-shaped and lies on the lingual face of the mandible wedged into the posterior end of the dentary. Posterodorsally, the splenial contacts the coronoid and the surangular; posteroventrally it is bounded by the angular. The relative size of the splenial is variable in iguanines, with that of *Sauromalus* being smaller than those of the other genera. Although there is considerable variation in the size of the splenial among the four outgroups used in

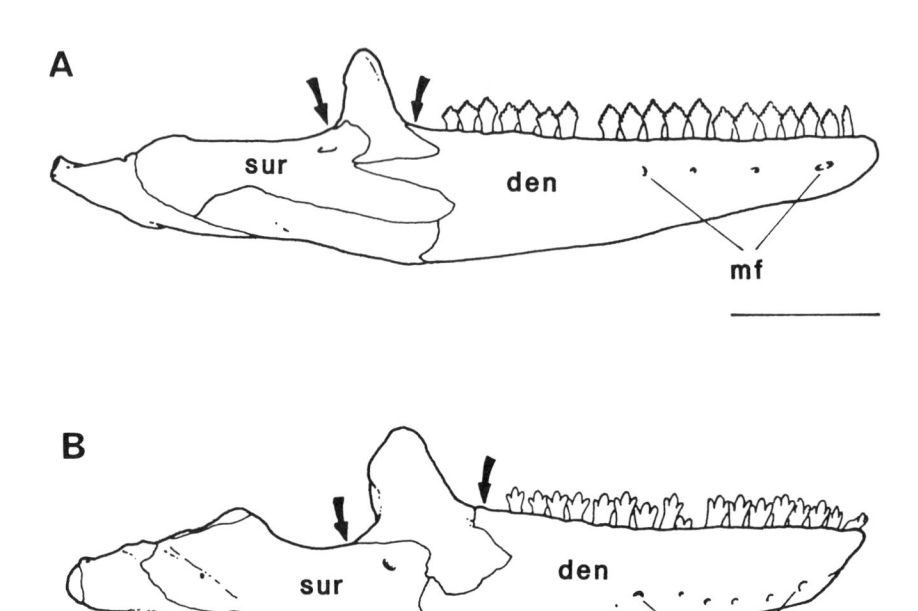

FIG. 22. Lateral views of the right mandibles of (A) *Iguana delicatissima* (MCZ 60823) and (B) *Amblyrhynchus cristatus* (RE 1396), showing differences in the relative heights of the dentary (den) and surangular (sur) and in the position of the row of mental foramina (mf). Scale equals 1 cm.

this study, all have a relatively larger splenial than *Sauromalus*. Therefore, I consider a small splenial to be apomorphic for iguanines.

The anterior inferior alveolar foramen pierces the mandible on its lingual surface at a point between one-third and one-half the way back from the anterior end of the jaw (Fig. 23). In most iguanines, this foramen lies within the suture between the splenial and the dentary at the anterior end of the splenial or along its anterodorsal edge. The coronoid may extend anteriorly between splenial and dentary so that it forms the posterior margin of the anterior inferior alveolar foramen (Fig. 23A) in some *Brachylophus*, *Dipsosaurus*, and *Sauromalus*. Varying amounts of this anterior extension of the coronoid may be covered by the splenial lingually, excluding the coronoid from the border of the foramen (Fig. 23B). This condition occurs in *Conolophus*, *Ctenosaura*, *Iguana*, most *Cyclura*, and in some *Brachylophus*, *Dipsosaurus*, and *Sauromalus*. In *Brachylophus*, the splenial is truncated, and the anterior inferior alveolar foramen sometimes lies entirely within the dentary. In *Amblyrhynchus*, the coronoid extends far anteriorly, and the foramen lies between it, rather than the dentary, and the splenial (Fig. 23C).

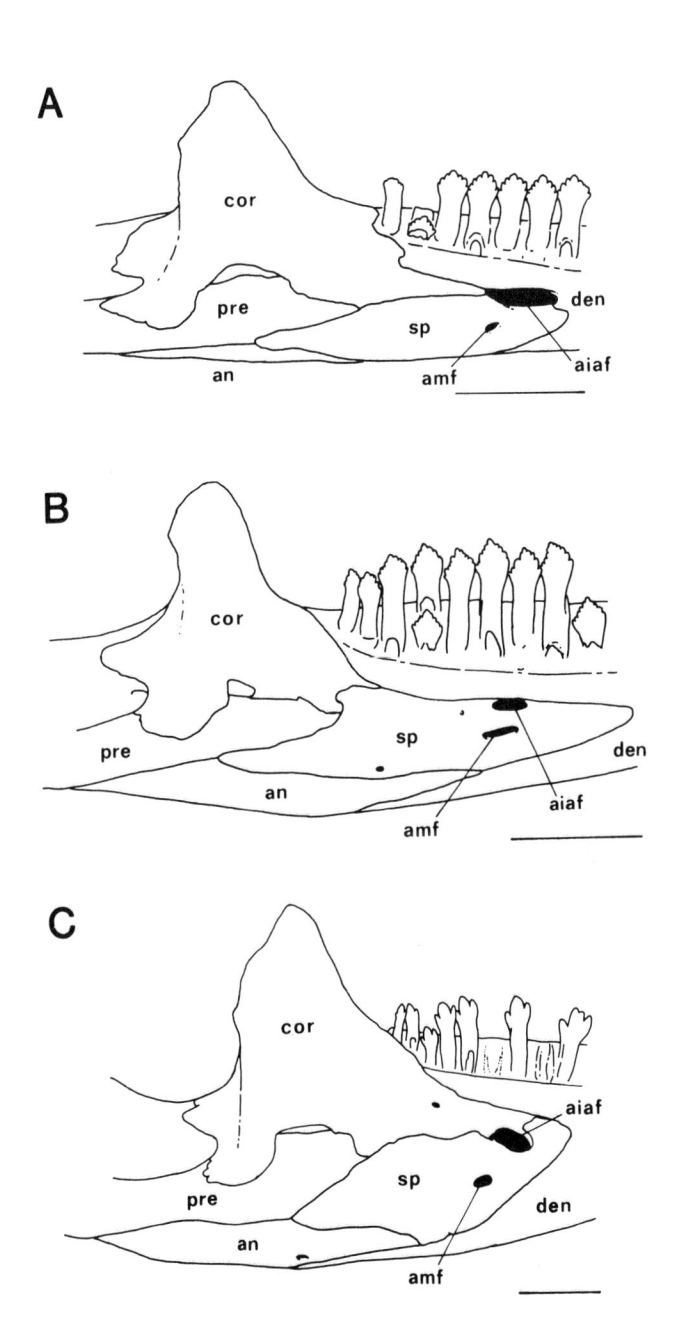

FIG. 23. Lingual views of the left mandibles of (A) *Sauromalus varius* (RE 512), (B) *Iguana delicatissima* (MCZ 60823), and (C) *Amblyrhynchus cristatus* (RE 1091), showing differences in the bones that surround the anterior inferior alveolar foramen. Scale equals 0.5 cm. Abbreviations: aiaf, anterior inferior alveolar foramen; amf, anterior mylohyoid foramen; an, angular; cor, coronoid; den, dentary; pre, prearticular; sp, splenial.

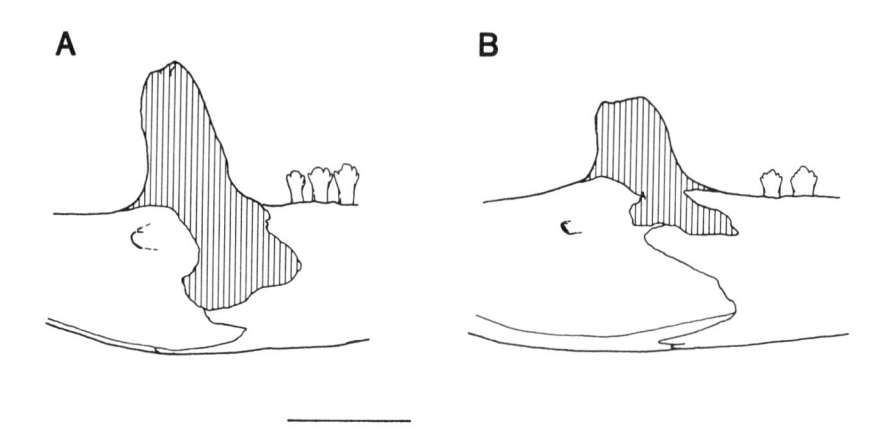

FIG. 24. Lateral views of the right mandibles of (A) *Conolophus pallidus* (RE 1382) and (B) *Cyclura cornuta* (RE 383), showing differences in the size of the labial process of the coronoid (shaded). Scale equals 1 cm.

Basiliscines, crotaphytines, morunasaurs, and oplurines have their anterior inferior alveolar foramina either between splenial and dentary or entirely within the splenial. Both conditions are found in all four outgroups. The splenial is relatively larger in most of these outgroups than in any iguanine, which may account for the fact that the foramen of iguanines does not lie entirely within this bone. Because location of the anterior inferior alveolar foramen between splenial and dentary is the only condition that occurs in both ingroup and outgroups, I considered this to be the plesiomorphic condition. The other two positions of the foramen, entirely within the dentary and between coronoid and splenial, were considered to be separate modifications of the plesiomorphic condition.

Coronoid (Figs. 6B,C, 23, 24). This bone forms a large dorsal process (coronoid eminence) immediately posterior to the tooth row, which serves as the insertion for jaw adductor muscles. It also bears one lateral and two medial ventrally directed processes that straddle the body of the lower jaw. Although absent in many iguanids, the large process of the coronoid that extends over the labial surface of the mandible is present in all iguanines (Fig. 24). This labial extension of the coronoid is most strongly developed in adult *Conolophus*, in which its ventral border reaches halfway or farther down the mandible and covers the posterolateral end of the dentary (Fig. 24A). In most other iguanines, the labial process of the coronoid is relatively small (Fig. 24B), but in *Amblyrhynchus* and *Brachylophus* the size of the process is intermediate between that of *Conolophus* and those of other iguanines. In both *Amblyrhynchus* and *Conolophus* the labial process of the coronoid is relatively small at hatching and increases in size during postembryonic ontogeny. The labial process of the coronoid is very small in basiliscines, crotaphytines, and oplurines. Morunasaurs and other iguanids that possess a large labial process of the

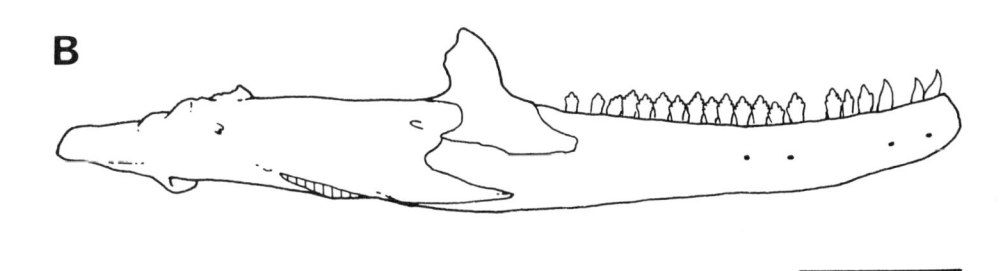

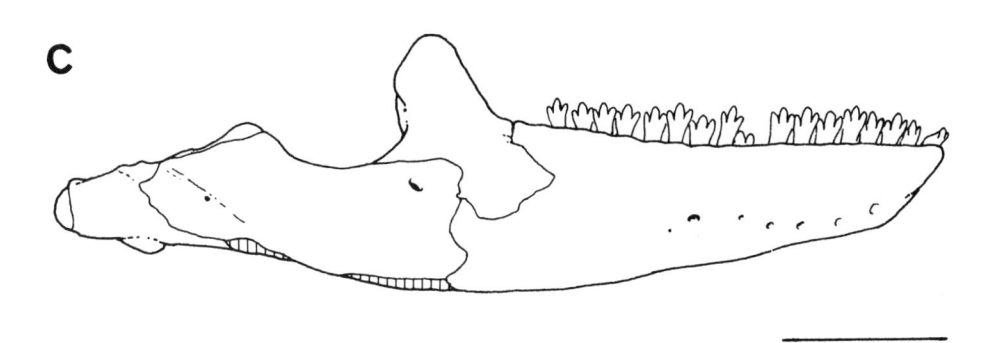

FIG. 25. Laterial views of the right mandibles of (A) *Iguana delicatissima* (MCZ 60823), (B) *Sauromalus obesus* (RE 467), and (C) *Amblyrhynchus cristatus* (RE 1396), showing differences in the lateral exposure of the angular (shaded). Scale equals 1 cm.

coronoid have a relatively slight ventral extension of this process compared to *Amblyrhynchus*, *Brachylophus*, and especially *Conolophus*.

Angular (Fig. 6B,C, 25). The angular is located on the ventral surface of the mandible, forming sutures with the splenial anterodorsally and the prearticular posterodorsally on the lingual surface of the mandible and with the dentary anteriorly and the surangular posteriorly on the labial side. In *Brachylophus*, *Ctenosaura*, *Cyclura*, *Dipsosaurus*, and *Iguana*, the angular extends far up the labial surface of the mandible so that it is easily seen in lateral view (Fig. 25A). The angulars of *Amblyrhynchus*, *Conolophus*, and *Sauromalus* are restricted labially so that they are barely visible from the lateral side (Fig. 25B,C). Compared to those of other iguanines, the angular of *Sauromalus* is relatively narrow. Because the angulars of basiliscines, crotaphytines, morunasaurs, and most oplurines are wide posteriorly and extend far up the labial surface of the mandible, I considered these to be plesiomorphic conditions. In *Oplurus*, the width and labial exposure of the angular are variable owing to varying degrees of reduction in this bone.

Surangular (Fig. 6B,C, 26, 27). This bone forms the dorsal portion of the mandible posterior to the coronoid and anterior to the articular facet. It fuses with the prearticular late in ontogeny. Dorsal to its suture with the angular on the labial surface of the jaw, the anterior extent of the iguanine surangular is variable (Fig. 26). In *Amblyrhynchus*, *Brachylophus*, and *Dipsosaurus* the exposed part of the surangular barely extends to the level of the apex of the coronoid, being covered by the dentary anterior to this level (Fig. 26A,B). In *Conolophus*, it extends slightly farther, to the level of the anterior slope of the coronoid eminence. The surangulars of *Iguana* and *Cyclura* extend far forward, well beyond the anterior slope of the coronoid eminence and often anterior to several of the posteriormost dentary teeth (Fig. 26C). *Sauromalus* and *Ctenosaura* are intermediate and variable within species; the surangular in each of these genera usually extends beyond the anterior slope of the coronoid eminence, but falls short of the tooth row. Some members of both genera exhibit a condition similar to that of *Conolophus*, and some *Ctenosaura* have a surangular that extends beyond the posteriormost dentary tooth.

Although the outgroups used in this study are also variable in the anterior extent of the surangular, in none does it extend as far forward as in *Iguana* and *Cyclura*. Therefore, in the absence of other information, it seems that a great anterior extent of the surangular is a synapomorphy of these two taxa. If the basic taxa used in this study are monophyletic, then a similar condition seen in some *Ctenosaura* must either be convergent, or the character may have arisen initially as a polymorphism, or some *Ctenosaura* have reverted to the ancestral morphology.

On the lingual side of the mandible, ventral to the apex of the coronoid in the arch between the ventral feet of this bone, a small portion of the surangular is variably visible in iguanines (Fig. 27). In most iguanines, this part of the surangular is relatively large and has the shape of a dome above the prearticular (Fig. 27A). In *Amblyrhynchus*, *Conolophus*, and *Cyclura cychlura*, the prearticular extends further dorsally, either completely excluding the surangular from the lingual surface of the mandible (Fig. 27B) or leaving only a thin sliver of it exposed. Although few small specimens were examined,

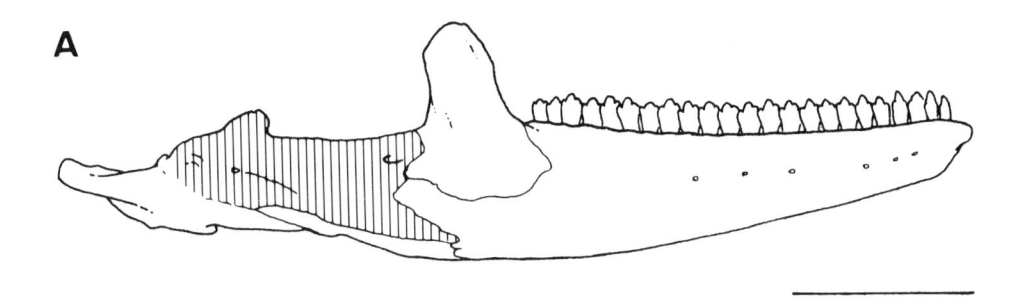

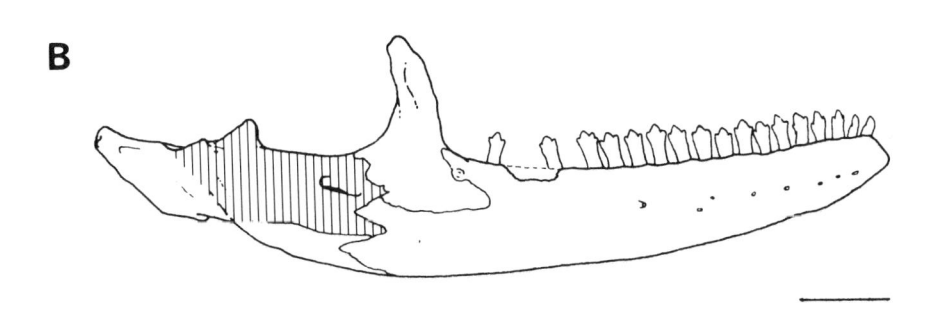

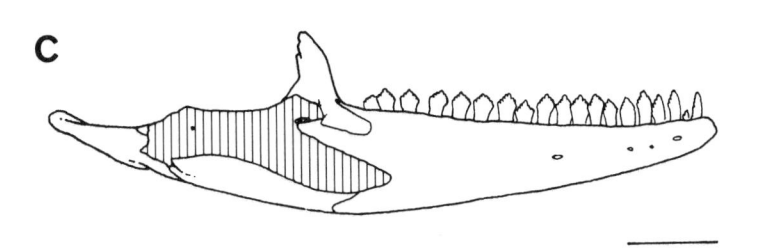

FIG. 26. Lateral views of the right mandibles of (A) *Dipsosaurus dorsalis* (RE 359), (B) *Brachylophus vitiensis* (MCZ 160254), and (C) *Iguana iguana* (RE 453), showing differences in the anterior extent of the surangular (shaded). Scale equals 0.5 cm.

there appears to be a transformation of this part of the surangular from exposed to unexposed during the postembryonic ontogenies of *Amblyrhynchus* and *Conolophus*. Some intraspecific variation exists in this feature; but other than the taxa in which the unexposed portion of the surangular is the common condition, only in *Brachylophus*

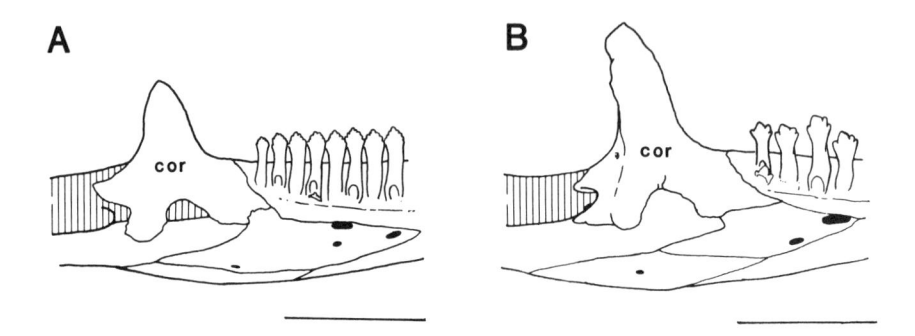

FIG. 27. Medial views of the left mandibles of (A) *Iguana delicatissima* (MCZ 16157) and (B) *Conolophus subcristatus* (MVZ 77314), showing differences in the exposure of the surangular (shaded) below the coronoid (cor). Scale equals 1 cm.

fasciatus, Cyclura nubila, and *Sauromalus varius* does this condition appear to be more than a rare variant.

Except for *Corytophanes* and *Oplurus quadrimaculatus,* all outgroups examined have a relatively large, dome-shaped portion of the surangular visible lingually between the ventral feet of the coronoid. In *Corytophanes,* however, lingual restriction of the surangular results from ventral extension of the coronoid rather than dorsal extension of the prearticular, the condition in iguanines. For this reason, as well as the hypothesis that *Basiliscus* rather than *Corytophanes* is the sister group of the other two basiliscine genera (Etheridge and de Queiroz, 1988), I considered the superficially similar conditions seen in *Corytophanes* and in some iguanines to be nonhomologous. Thus, the large lingual exposure of the surangular between coronoid and prearticular is interpreted as plesiomorphic.

Prearticular (Figs. 6B,C, 28, 29). This bone forms the ventromedial portion of the posterior end of the mandible. The prearticular bears two processes for the insertion of jaw adductor and abductor muscles, the posteriorly directed retroarticular process and the medially directed angular process. The retroarticular process is large in all iguanines, but the relative size of the angular process is variable. In all iguanines except *Amblyrhynchus,* the angular process is small at hatching and increases in relative size as the animal grows (Fig. 28A-C). The angular process of *Amblyrhynchus* is very small in juveniles and increases in relative size only slightly during postembryonic ontogeny (Fig. 28D-F); even in large adults it has only about the same relative size as those of young of other iguanine genera.

Except for *Corytophanes* and *Laemanctus,* all outgroup taxa examined (including those that are small as adults) have relatively large angular processes. Thus, if basiliscines are the sister group of iguanines, then the polarity of this character is equivocal; if not, then the development of a large angular process during ontogeny must be considered to be plesiomorphic. Because *Amblyrhynchus* exhibits the nontransforming ontogeny, strict

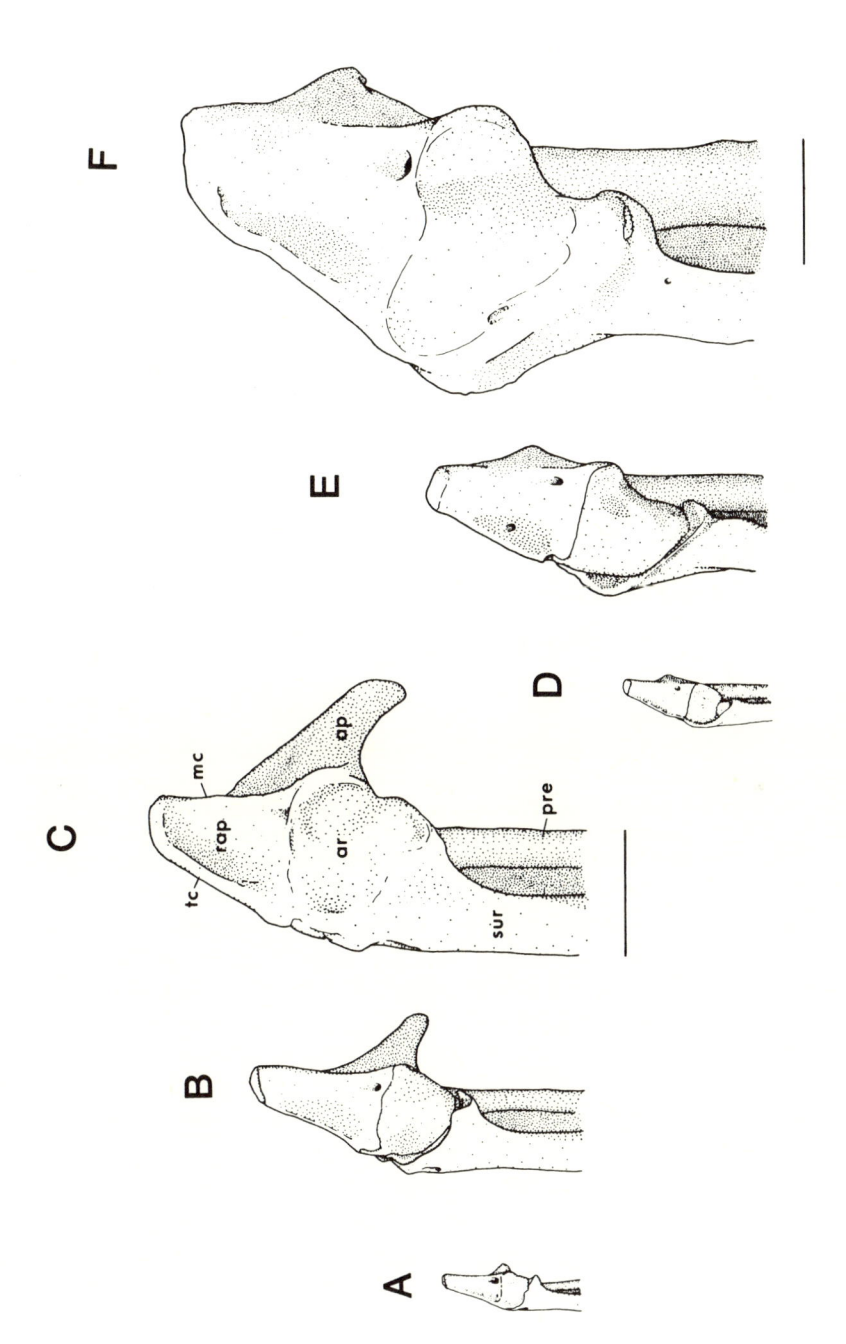

FIG. 28. Dorsal views of the posterior ends of the right mandibles of (A-C) three *Ctenosaura hemilopha*-(A) RE 1386, (B) RE 498, (C) RE 325-and (D-F) three *Amblyrhynchus cristatus*-(D) SDNHM 45156, (E) RE 2239, (F) RE 1095-showing differences in the ontogeny of the angular process and the distinctness of the medial crest of the retroarticular process. Scale equals 0.5 cm. Abbreviations: ap, angular process; ar, articular; mc, medial crest; pre, prearticular; rap, retroarticular process; sur, surangular; tc, tympanic crest.

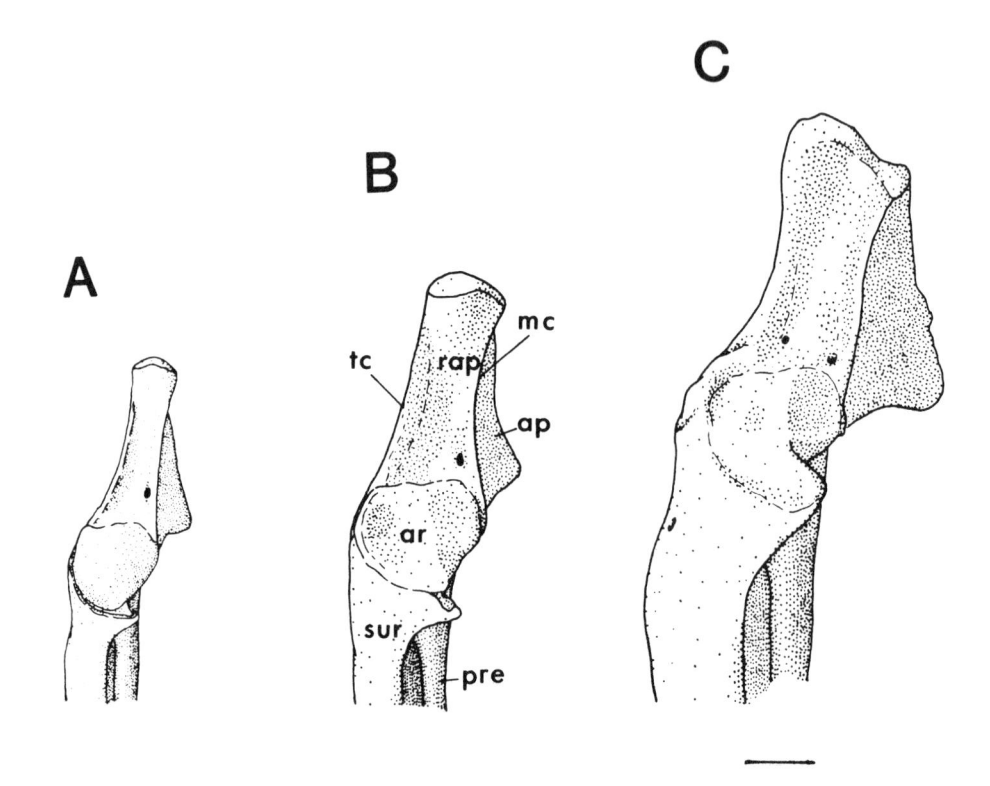

FIG. 29. Dorsal views of the posterior ends of the right mandibles of three *Dipsosaurus dorsalis*-(A) RE 355, (B) RE 1576, (C) RE 356-showing ontogenetic divergence of the posterior ends of the medial and tympanic crests of the retroarticular process. Scale equals 1 mm. Abbreviations: ap, angular process; ar, articular; mc, medial crest; pre, prearticular; rap, retroarticular process; sur, surangular; tc, tympanic crest.

adherence to the ontogenetic criterion for assessing character polarity (Nelson and Platnick, 1981; Patterson, 1982) would, in this case, produce results that conflict with those of outgroup comparison. Phylogeny, however, is a procession of changing ontogenies. Therefore, I treat the ontogenetic development of a large angular process and the lack of such development as different character states. When characters are conceived as ontogenetic transformations rather than instantaneous forms (e.g., large vs. small angular process), there can be no conflict between ontogeny and outgroups. Ontogenetic transformations do not provide information about character phylogeny; they are the characters themselves (de Queiroz, 1985).

The retroarticular process of iguanines extends from the posterior end of the mandible. It is bounded on either side by crests: the tympanic crest laterally, and medially by what I will simply call the medial crest. In most iguanines, the two crests converge posteriorly, and the retroarticular process is triangular (Fig. 28). This condition is also seen in juvenile

Dipsosaurus, but in this taxon the posterior ends of the crests move apart during ontogeny so that the retroarticular process of large *Dipsosaurus* is quadrangular (Fig. 29).

Most outgroups have a triangular retroarticular process, much like those seen in the majority of iguanines; however, I have observed quadrangular retroarticular processes in *Morunasaurus annularis* and *Enyalioides praestabilis*. Thus, either the quadrangular retroarticular process of *Dipsosaurus* is apomorphic or the polarity of this character is equivocal, but a quadrangular retroarticular process will never be considered to be plesiomorphic with the outgroups used in this study.

The medial crest of the retroarticular process varies in size within Iguaninae. In *Amblyrhynchus* (Fig. 28D-F), *Brachylophus*, *Conolophus*, and *Cyclura cornuta*, this structure is but a low, rounded ridge, contrasting with the sharp crest seen in other iguanines (Figs. 28A-C, 29). Intraspecific variability in *Amblyrhynchus* and *Conolophus*, but more important, variation within basiliscines, morunasaurs, and oplurines, prevented me from using the size of the medial crest as a character for phylogenetic analysis.

Articular (Figs. 6C, 28, 29). The articular bone is the ossified posterior end of Meckel's cartilage and forms the condyle that articulates with the quadrate of the skull proper. It sits in a groove in the dorsal surface of the jaw between the prearticular posteriorly and medially and the surangular anterolaterally. The articular of iguanines fuses to the prearticular around the time of hatching. I have not studied variation in the iguanine articular.

MISCELLANEOUS HEAD SKELETON

Marginal Teeth (Figs. 5B, 8, 30). The marginal teeth of iguanines exhibit a bewildering diversity of form and could easily be the subject of a study by themselves. Some dentitional features common to all iguanines are pleurodonty and the formation of replacement teeth directly lingual to the teeth being replaced (iguanid tooth-replacement pattern of Edmund, 1960). Although lizards are often stereotyped as being homodont, all iguanines exhibit some regional differentiation in the morphology of their marginal teeth. This differentiation is most pronounced, at least in terms of crown morphology, in *Cyclura* and *Sauromalus*, where the crowns of the anterior teeth are conical and usually lack lateral cusps while those of the posterior teeth are laterally compressed and polycuspate. Another feature common to all iguanines is an allometric increase in tooth number within species, a feature that has been reported previously in iguanines (Ray, 1965; Montanucci, 1968) and in various other iguanids (Etheridge, 1962, 1964b, 1965a; Ray, 1965). This allometric increase in tooth number results from the addition of teeth to the posterior ends of the maxillary and dentary tooth rows; the number of premaxillary teeth remains constant.

Variation in the number of premaxillary teeth of iguanines is given in Table 3. Most or all species of *Amblyrhynchus*, *Brachylophus*, *Conolophus*, *Ctenosaura*, *Dipsosaurus*, and *Iguana* have a statistical mode of seven premaxillary teeth. The species of *Cyclura* generally have modes of greater than seven premaxillary teeth, and those of *Sauromalus* have modal numbers lower than seven. *Ctenosaura defensor* also has fewer than seven

TABLE 3. Numbers of Premaxillary Teeth

Taxon	N	\multicolumn{9}{c}{Number of Premaxillary Teeth}
		3	4	5	6	7	8	9	10	11
Amblyrhynchus cristatus	16			6%	6%	*88%*				
Brachylophus fasciatus	15				7	*93*				
vitiensis	2					*50*		*50%*		
Conolophus pallidus	14					*100*				
subcristatus	11					*100*				
Ctenosaura acanthura	17				6	*70*	12%	6		
bakeri	12				25	*75*				
clarki	9			11	11	*67*		11		
defensor	5			*80*	20					
hemilopha	15					*100*				
palearis	8					*100*				
pectinata	12					*92*	8			
quinquecarinata	11			9	18	*73*				
similis	18				11	*67*	17	6		
Cyclura carinata	5				20	20	*40*	20		
collei	1						*100*			
cornuta	8					13	*38*	50		
cychlura	7						29	*71*		
nubila	8						38	*38*	12%	12%
pinguis[1]	2					50	*50*			
ricordii	4						25	*75*		
rileyi	8					38	*50*	12		
Dipsosaurus dorsalis	22				23	*73*	5			
Iguana delicatissima	6					*100*				
iguana	21				19	*76*	5			
Sauromalus ater	6		*67%*	17	7					
australis	2			*100*						
hispidus	15			*40*	40	20				
obesus	19	5%	10	*58*	21	5				
slevini	2			*100*						
varius	15		7	*47*	47					

Note: The figures given under the different numbers of premaxillary teeth are percentages. Modes are in italics.
[1] Additional data from Pregill (1981).

premaxillary teeth. Two specimens of *Cyclura pinguis* have seven and eight premaxillary teeth. I have assumed that *C. pinguis* actually has a modal number of premaxillary teeth greater than seven and that the bimodal distribution results from sampling error. It is also possible that a phylogenetic transformation has occurred within *Cyclura* and that the synapomorphic condition applies to a subset of this taxon, or that the ancestral condition was polymorphic.

Outgroup comparison yields equivocal results concerning the plesiomorphic number of premaxillary teeth in iguanines. *Gambelia* has the condition found in most iguanines, a mode of seven premaxillary teeth. Other outgroups have seven or more premaxillary teeth (basiliscines, *Enyalioides*); more than seven (*Morunasaurus*); fewer than seven (oplurines, *Hoplocercus*); or a range from fewer than seven to more than seven (*Crotaphytus*, mode of six). Because of this ambiguity, I withheld a decision on the primitive number of premaxillary teeth and used the character only at a level less inclusive than all iguanines.

In most iguanines the premaxillary teeth, as well as the anterior maxillary and dentary teeth, have fewer or smaller cusps than the posterior maxillary and dentary teeth. In *Cyclura* and most species of *Ctenosaura* the premaxillary teeth and the dentary teeth with which they occlude lack lateral cusps. At least some of the premaxillary teeth of some specimens have one or more lateral cusps in *Brachylophus*, *Dipsosaurus*, and *Ctenosaura palearis*, although these lateral cusps are relatively small. *Amblyrhynchus* and *Conolophus* almost invariably have two large lateral cusps on their premaxillary teeth. The premaxillary teeth of basiliscines, crotaphytines, morunasaurs, and oplurines usually lack lateral cusps, though small ones may occasionally be present.

Except in large *Ctenosaura*, in which the anterior maxillary teeth and the dentary teeth occluding with them are enlarged and recurved to form fangs, these teeth differ only slightly from the marginal teeth anterior to them. Moving posteriorly along the marginal tooth rows, the tooth crowns progressively become more laterally compressed, the size of the lateral cusps increases, and in most iguanines additional lateral cusps are added. Part of the progression is reversed abruptly at the posterior ends of the tooth rows. When strongly compressed, the crowns of the teeth are much wider than their bases and overlap their neighbors in a regular pattern: each tooth is twisted about its long axis so that its anterior edge is lingual to and its posterior edge is labial to the crowns of the adjacent teeth. Maximum cuspation is reached about three-fourths of the way back along the tooth row in adults, and here substantial differences exist among taxa (Fig. 30). The maximum number of cusps on the marginal teeth of *Brachylophus*, *Conolophus*, *Dipsosaurus*, and most *Ctenosaura* (*C. acanthura*, *C. clarki*, *C. hemilopha*, *C. palearis*, *C. pectinata*, and *C. similis*) is four: two anterior cusps, an apical cusp, and one posterior cusp (Fig. 30A). This crown morphology is seen in both maxillary and dentary teeth. The size and occurrence of the anteriormost cusp, however, is variable, and it may be absent from all teeth in some specimens of some species.

Greater cuspation is found in *Ctenosaura defensor*, *Cyclura*, and *Sauromalus* (Fig. 30B). The maximum number of cusps per tooth in these taxa ranges from as few as five in *Cyclura pinguis* and some *C. cychlura* up to about 10 in *C. cornuta* and *C. nubila*.

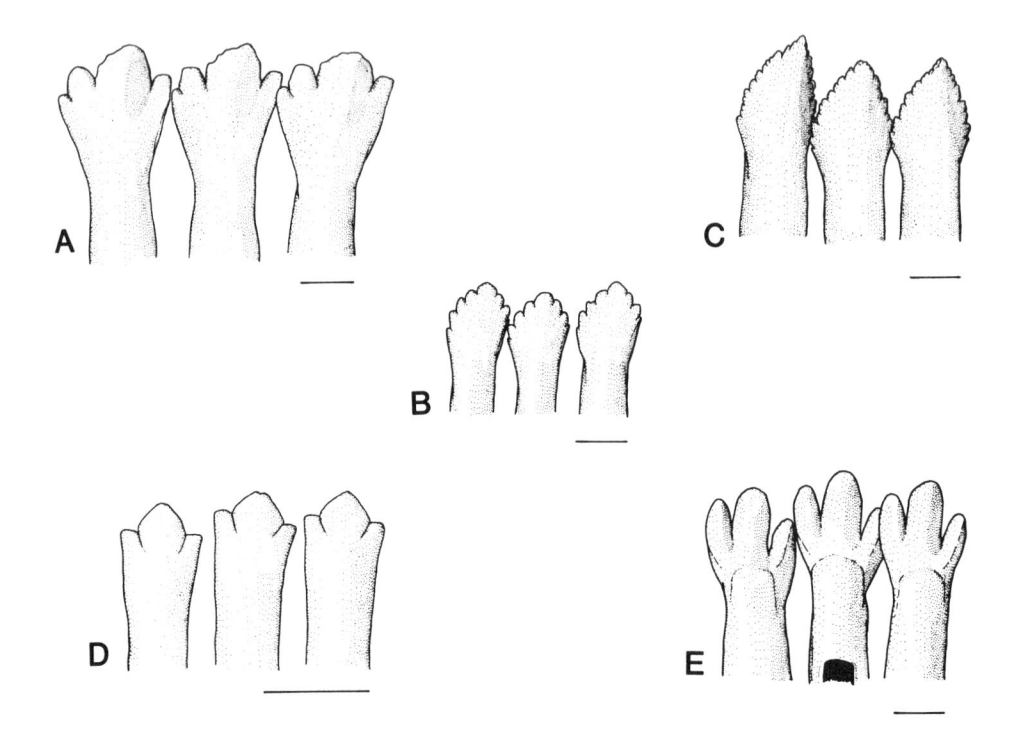

FIG. 30. Lingual views of left maxillary teeth of (A) *Conolophus pallidus* (RE 1382), (B) *Sauromalus varius* (RE 539), (C) *Iguana iguana* (JMS 1028), (D) *Basiliscus plumifrons* (RE 427), and (E) *Amblyrhynchus cristatus* (RE 1387), showing differences in cuspation. Scale equals 1 mm.

Increase in cuspation is accompanied by a difference in the morphology of the maxillary versus dentary teeth: maxillary teeth bear more cusps along their anterior edges, while cuspation of the dentary teeth is more or less symmetrical (Avery and Tanner, 1964:Fig. 3). Within the tooth row of a single organism, increase in cuspation appears to result from addition of cusps to the anterior and posterior edges of the crowns.

Still greater cuspation occurs in *Iguana*, reaching an extreme in *I. iguana*. In this genus the teeth possess a large number of small cusps, giving them a serrated cutting edge (Fig. 30C). The small cusps are difficult to count, especially when worn, but the maximum number is greater than 15 in *I. delicatissima* and greater than 20 in *I. iguana*. Cuspation increases both ontogenetically and from anterior to posterior in a single tooth row by two mechanisms: addition of cusps and subdivision of the fields of preexisting ones. The actual cusps of fully formed teeth are not subdivided, though their fields appear to be when teeth are compared with their replacements; it is, of course, impossible to have actual subdivision of cusps from one tooth to the next. Because cuspation increases ontogenetically, the teeth of young *Iguana* have about as many cusps as do those of some large *Cyclura*. The maximum number of cusps in mature *Iguana*, however, is greater than in any *Cyclura*.

Amblyrhynchus, *Ctenosaura bakeri*, and *C. quinquecarinata* are the only iguanines that characteristically have a maximum of only three cusps on their marginal teeth. Tricuspid teeth occur throughout the posterior half of the tooth row in juveniles of at least some iguanine species whose teeth later become four-cusped or polycuspate, and they are common outside of iguanines, occurring in basiliscines (Fig. 30D), crotaphytines, oplurines, and most morunasaurs (some *Enyalioides* are polycuspate). For these reasons, tricuspid posterior marginal teeth are judged to be plesiomorphic for iguanines. The morphology of the tooth crowns in the outgroups, however, differs strikingly from that of *Amblyrhynchus*, although it is similar to that of the tricuspid teeth found more anteriorly in the tooth row or earlier in the ontogeny of other iguanines. In the tricuspid teeth of all these taxa, the apical cusp is much larger than each lateral cusp. In *Amblyrhynchus*, the lateral cusps are very large, each being nearly as large as the apical cusp (Fig. 30E). The posterior marginal teeth of *Ctenosaura quinquecarinata* are similar to those seen in many outgroup taxa.

Ontogenetic data relating to changes in iguanine tooth crown morphology are few, but what little are available suggest that the adult morphologies of the marginal tooth crowns represent stages in a single transformation series. Tricuspid teeth are judged to be plesiomorphic on the basis of outgroup comparison (see above), and they also occur in the few hatchling specimens examined of those iguanines that, as adults, have four-cusped teeth (*Conolophus subcristatus*, *Ctenosaura hemilopha*, *C. pectinata*, *C. similis*), polycuspate teeth (*Cyclura carinata*, *C. cornuta*, *C. nubila*), and serrate teeth (*Iguana iguana*) as adults. Although I have never observed the replacement of four-cusped teeth by polycuspate or serrate teeth, both *Sauromalus* and *Cyclura* (which are polycuspate as adults) normally possess four-cusped teeth in some portion of the tooth row. Thus, all iguanine tooth crown morphologies appear to be part of a single transformation series, with tricuspid teeth in the terminal stage at its plesiomorphic pole. I also propose that ontogenetic transformation to polycuspate teeth is a modification of a transformation to four-cusped teeth, and that ontogenetic transformation to serrate teeth is a modification of one to polycuspate teeth.

Judging from the high numbers of replacement teeth in *Amblyrhynchus*, these animals probably replace their teeth at higher rates than other iguanines and the members of the four noniguanine outgroups examined in this study. Presumably related to the high numbers of replacement teeth in *Amblyrhynchus* is a relatively wide alveolar margin on the bones bearing the marginal teeth.

Palatal Teeth (Fig. 31). Palatal teeth in iguanids may be present on the pterygoids and palatines but never on the vomers. All iguanines lack palatine teeth, which are present (though not invariably) in crotaphytines and oplurines among the outgroups examined. At least some specimens of all iguanine species examined in this study have pterygoid teeth, the number and position of which vary considerably among genera. The pterygoid teeth generally lack lateral cusps (in contrast with the tricuspid pterygoid teeth of some basiliscines) and are directed posteroventrally; the tips of these teeth may also curve posteriorly. In most iguanines, the number of pterygoid teeth increases ontogenetically, though this increase is less conspicuous in species with small maximum numbers of pterygoid teeth.

Pterygoid teeth are present in all four outgroups examined and lie in a single row close to the ventromedial edge of each pterygoid, next to the pyriform recess. The posterior end of the row may be displaced slightly laterally. This plesiomorphic condition is retained in *Brachylophus*, and is also seen in some *Cyclura* and *Sauromalus* as an individual variant. A modification of this condition seems to have occurred by lateral displacement of the posterior end of the tooth row toward the base of the transverse process of the pterygoid, with an accompanying tendency for this posterior portion of the tooth row to double ontogenetically. Beneath the posterior end of the tooth row a bony mound may be raised. An ontogenetic transformation from the presumed plesiomorphic condition mirrors the hypothesized phylogenetic transformation of terminal morphologies based on outgroup comparison. This apomorphic condition is seen in adult *Ctenosaura* and in some *Cyclura* and *Sauromalus*.

Two independent phylogenetic transformations appear to have been derived from the apomorphic condition described above. The first, seen in *Iguana*, results ontogenetically and presumably was derived phylogenetically from an increase in the number of pterygoid teeth and a more extensive doubling of the tooth row late in ontogeny. The second, seen in *Amblyrhynchus*, apparently resulted from loss of the anterior portion of the tooth row; the remaining teeth are located in a short, laterally displaced patch, even in juveniles.

Pterygoid teeth are usually absent in *Conolophus* and *Dipsosaurus* (occasionally absent in individual specimens of *Sauromalus*), but their absence in these two taxa appears to represent separate derivations from different antecedent conditions. In the rare specimens of *Dipsosaurus* that have pterygoid teeth, these teeth are present in a single row near the medial edge of the bone, suggesting derivation from the plesiomorphic condition. This inference is complicated by the small size of *Dipsosaurus* combined with the large size at which lateral displacement of the row occurs in taxa that exhibit this derived condition. When pterygoid teeth are present in *Conolophus* they are located laterally, near the base of the transverse process. This suggests that lateral displacement of the posterior end of the tooth row (an apomorphic condition) preceded tooth loss; the reduction of the anterior end of the tooth row seen in *Amblyrhynchus* is a likely intermediate state.

Figure 31 is a hypothetical character phylogeny for the iguanine pterygoid tooth patch. The three most speciose iguanine genera, *Ctenosaura*, *Cyclura*, and *Sauromalus*, exhibit much variation in their pterygoid teeth. They are all considered to exhibit one of the two initial modifications of the plesiomorphic condition in the diagram, although this treatment ignores much of the actual variation. Because of the complexity of this character, it is necessary to subdivide it into three characters so that coding will accurately reflect the hypothesized phylogenetic transformations.

The number of teeth on a single pterygoid is highly variable among iguanine taxa; however, allometric increase in this feature makes intertaxic comparison difficult among

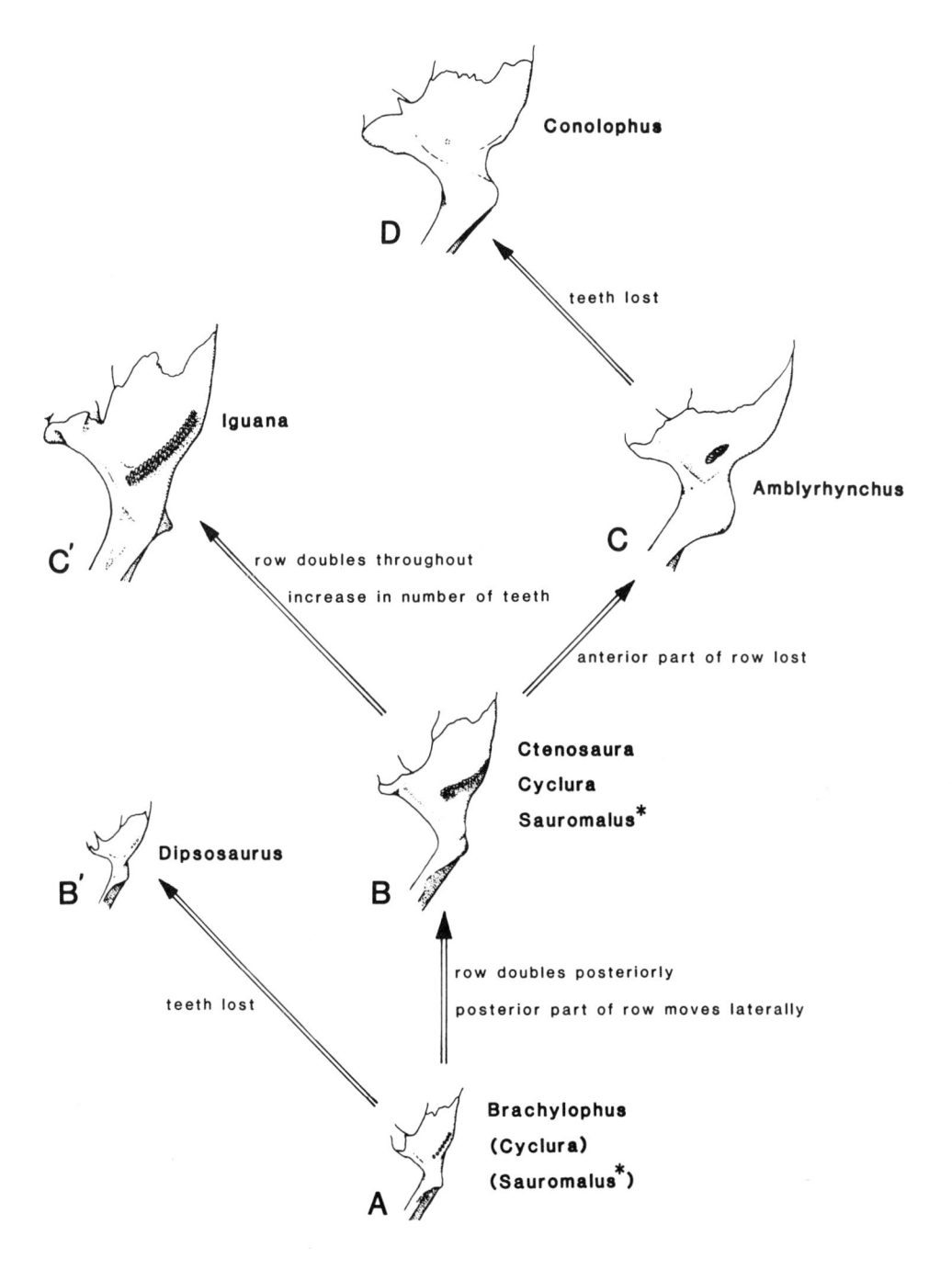

FIG. 31. Hypothetical character phylogeny for the iguanine pterygoid tooth patch. An asterisk indicates that pterygoid teeth are sometimes absent; parentheses indicate a rare condition in the enclosed taxon. See text for details.

taxa whose organisms reach different sizes. As stated above, *Conolophus* and *Dipsosaurus* generally lack pterygoid teeth, although I have observed up to two and four on a single pterygoid in these genera, respectively. *Amblyrhynchus*, *Brachylophus*, and *Sauromalus* generally have fewer than 10 pterygoid teeth and always have less than 15 (the maximum numbers that I have observed are seven, 11, and 12, respectively). Because of the wide range in body size of their included species, *Ctenosaura* and *Cyclura* exhibit a wide range in pterygoid tooth number. Members of the large species of *Ctenosaura* (*C. acanthura*, *C. pectinata*, *C. similis*) usually have over 20 pterygoid teeth and sometimes exceed 30. Small species such as *C. clarki*, *C. defensor*, *C. palearis*, and *C. quinquecarinata* probably never have as many as 20 such teeth. *Cyclura* exhibits a range in the number of pterygoid teeth similar to that of *Ctenosaura*, but I have few adequate ontogenetic series for species in the former genus. The most teeth that I have seen on a single pterygoid in *Cyclura* is 26 in a specimen of *C. pingius* that had not yet undergone the fusion of braincase elements indicative of the attainment of maximum size. If allometric trends in this species are similar to those in *Iguana* and *Ctenosaura*, larger organisms probably have upwards of 30 such teeth. *Iguana* is characterized by a high pterygoid tooth number. Large *I. delicatissima* have a maximum of at least 30 pterygoid teeth, while the number exceeds 60 in *I. iguana*. I did not use variation in pterygoid tooth number as a separate systematic character, though some of this variation is incorporated in the characters that were used.

Scleral Ossicles (Fig. 32). The scleral ossicles are thin wafers of bone that overlap one another in such a way that they form a ring within the sclera on the corneal side of the eye. The number of scleral ossicles and their pattern of overlap is fairly constant within squamate species, and a standard terminology has been developed to describe and number individual ossicles for purposes of comparison (Gugg, 1939; Underwood, 1970). Most Iguanidae characteristically possess 14 scleral ossicles per eye, with the following patterns of overlap: ossicles 1, 6, and 8 overlap both immediately adjacent ossicles; ossicles 4, 7, and 10 are overlapped by both immediately adjacent ossicles; and the remainder are overlapped by one neighboring ossicle while overlapping the other (Underwood, 1970; de Queiroz, 1982). In a previous study (de Queiroz, 1982), I reported this pattern for all iguanine genera. I have now examined the following additional species and report the same ossicle configuration: *Brachylophus vitiensis* (one eye from one specimen examined); *Ctenosaura bakeri* (Roatán Island; 2, 1); *C. clarki* (4, 4); *C. defensor* (1, 1); *C. palearis* (1, 1); *C. quinquecarinata* (1, 1); *C. similis* (8, 5); *Cyclura carinata* (4, 2); and *C. rileyi* (2, 1). Additional material of *Amblyrhynchus* (2, 1) also exhibits this pattern, supporting my previous suggestion that two specimens with fewer than 14 ossicles are anomalous.

Hyoid Apparatus (Fig. 33). The hyoid apparatus lies within the tissue between the mandibles, where it serves as the skeletal framework for the tongue and throat muscles. This delicate structure is often lost or partially destroyed in dry skeletal preparations. In iguanines, the hyoid apparatus consists of a median, anteriorly directed hypohyal (lingual process); the body of the hyoid, which is also a median element and is continuous with the hypohyal; and portions of three pairs of visceral arches. The hyoid arch is the most lateral

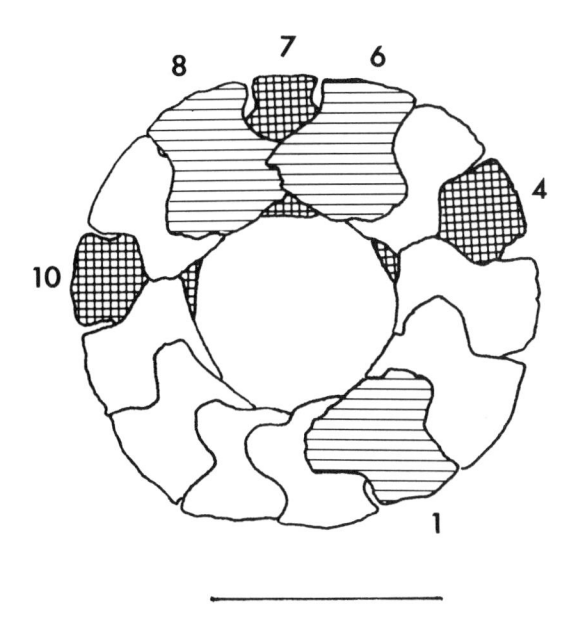

FIG. 32. Corneal view of the left scleral ring of *Ctenosaura similis* (MCZ 9566). All iguanine species typically exhibit the pattern of scleral ossicles illustrated: a total of 14 ossicles, with numbers 1, 6, and 8 positive (horizontal lines) and numbers 4, 7, and 10 negative (crosshatched). Scale equals 0.5 cm.

and consists of basihyals, projecting anterolaterally from the body of the hyoid, and ceratohyals, which run posteriorly from the distal ends of the basihyals. Basihyals fuse to the hyoid body late in postembryonic ontogeny. Separate epihyals are not evident. Medial to the ceratohyals lie the first ceratobranchials; these are the only bony elements of the hyoid apparatus, the remainder being composed of calcified cartilage. The first epibranchials extend posteriorly and dorsally from the posterior ends of the first ceratobranchials. The second ceratobranchials lie medial to the first ceratobranchials and extend directly posteriorly. Camp (1923) reported the presence of second epibranchials in *Iguana*. Although I have never observed discrete second epibranchials in iguanines, the delicate nature of these elements may have resulted in their destruction during skeletal preparation.

Differences exist among iguanine taxa in the relative lengths and the orientations of the various hyoid elements (Fig. 33). The most obvious differences are seen in the second ceratobranchials. In *Ctenosaura, Cyclura, Dipsosaurus,* and *Iguana delicatissima*, the second ceratobranchials are of moderate size; they are generally more than two-thirds the length of the first ceratobranchials, and never do they more than barely exceed the latter in length (Fig. 33A). Although there is some overlap in the ranges of the relative lengths of the second ceratobranchials between *Amblyrhynchus, Conolophus,* and *Sauromalus,* on the one hand, and members of the previously described group, the second ceratobranchials

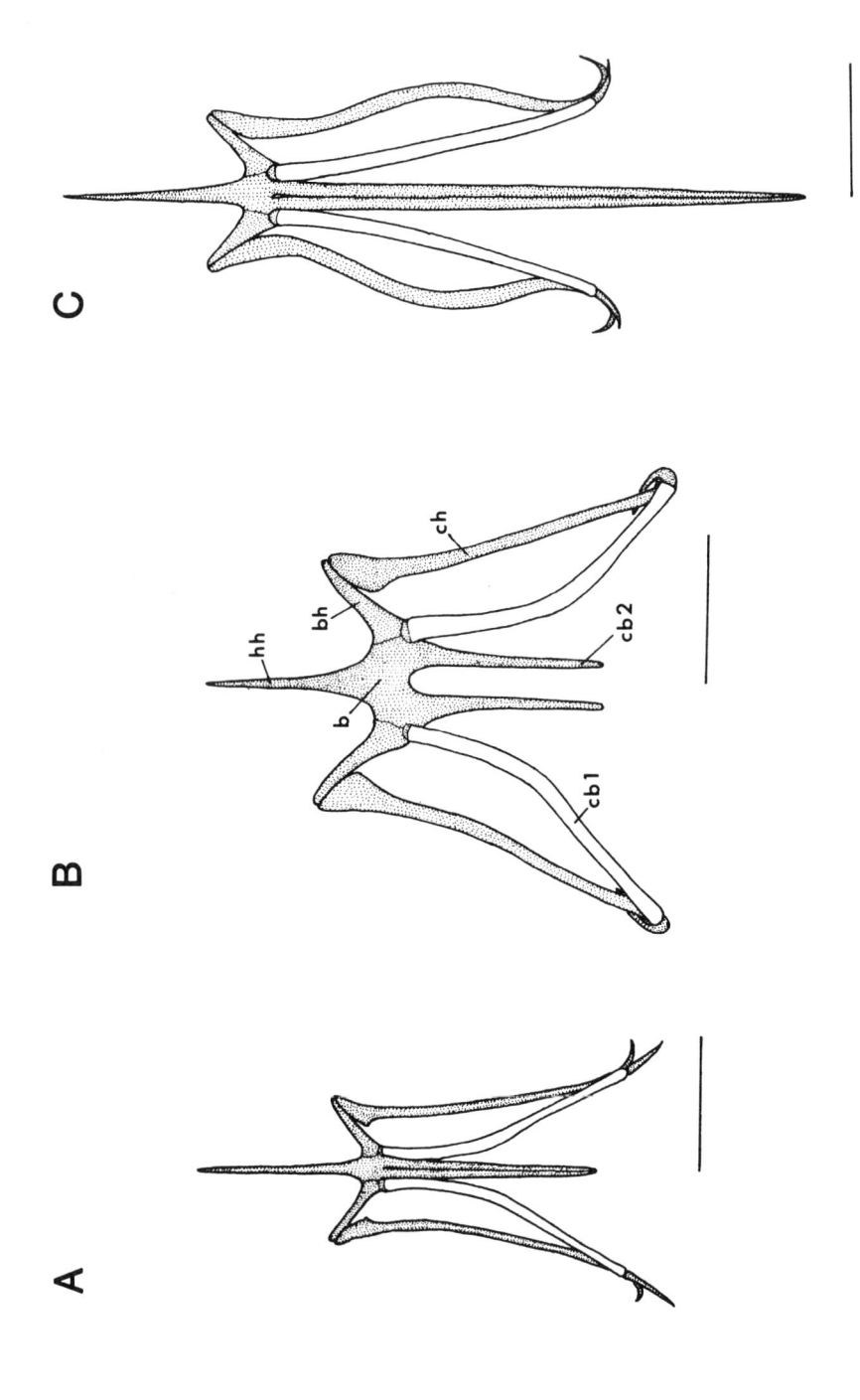

FIG. 33. Ventral views of the hyoid apparati of (A) *Ctenosaura quinquecarinata*, (B) *Sauromalus varius*, and (C) *Brachylophus vitiensis*, showing differences in the lengths of the second ceratobranchials and presence or absence of medial contact of these elements. Drawings are composite reconstructions. Scale equals 1 cm. Abbreviations: b, body of hyoid; bh, basihyal; cb1, first ceratobranchial; cb2, second ceratobranchial; ch, ceratohyal; hh, hypohyal.

of *Amblyrhynchus*, *Conolophus*, and *Sauromalus* are relatively short, often less than two-thirds the length of the first ceratobranchials (Fig. 33B). *Iguana iguana* and both species of *Brachylophus* have long second ceratobranchials, invariably much longer than the first ceratobranchials (Fig. 33C). The long second ceratobranchials support the gular fans seen in these species.

Another variable character in the hyoid skeletons of iguanines is the proximity of the two second ceratobranchials to one another. In all iguanines except *Amblyrhynchus* and *Sauromalus*, these elements contact each other along the midline for most or all of their lengths (Fig. 33A,C); sometimes they are separated by a small gap where they meet the body of the hyoid. In *Amblyrhynchus* and *Sauromalus* the second ceratobranchials are largely or entirely separated from one another (Fig. 33B).

Most of the outgroup taxa examined in this study have second ceratobranchials of intermediate size, these elements being slightly shorter than the first ceratobranchials. Some *Basiliscus* have slightly longer second ceratobranchials, but they are not nearly as long as those of *Brachylophus* and *Iguana iguana*. *Crotaphytus* and *Gambelia* have short second ceratobranchials, about half the length of their first ceratobranchials. Thus, very long second ceratobranchials are almost certainly apomorphic for iguanines, and, unless crotaphytines are the sister group of iguanines, short ones are probably also apomorphic. Separation of the second ceratobranchials along the midline is unequivocally apomorphic, based on the outgroups used in this study.

AXIAL SKELETON

Presacral Vertebrae (Figs. 34, 35, 36, 37). The presacral vertebrae (Fig. 34) of all iguanines are procoelous and possess supplementary articular surfaces, zygosphenes and zygantra, medial to the zygapophyses. Iguanine cervical vertebrae, defined as those vertebrae anterior to the first one bearing a rib that attaches to the sternum (Hoffstetter and Gasc, 1969) and including the atlas and axis, invariably number eight. From four to seven ventrally keeled intercentra are present on the atlas, the axis, and between the centra of the anterior cervical vertebrae, decreasing in size posteriorly. The intercentrum of the axis fuses with its centrum late in postembryonic ontogeny. There is regional differentiation in the shape of the presacral vertebrae: the anterior and posterior presacrals are relatively short compared to those in the middle of the column.

The number of presacral vertebrae in iguanines ranges from 23 to 27 (Table 4). Most species exhibit a strong statistical mode of 24 presacral vertebrae, with occasional variants having 23 or 25. I judge this to be the plesiomorphic condition because it is seen in all species of basiliscines, crotaphytines, morunasaurs, and oplurines that I have examined. Within the genus *Ctenosaura*, three species, *C. clarki*, *C. defensor*, and *C. quinquecarinata*, have a modal number of 25 presacral vertebrae. Because the apomorphic condition occurs in only some *Ctenosaura*, this character reveals nothing about relationships among my basic taxa. I used differences in modal numbers of presacral vertebrae as a character only in an analysis of relationships within *Ctenosaura*.

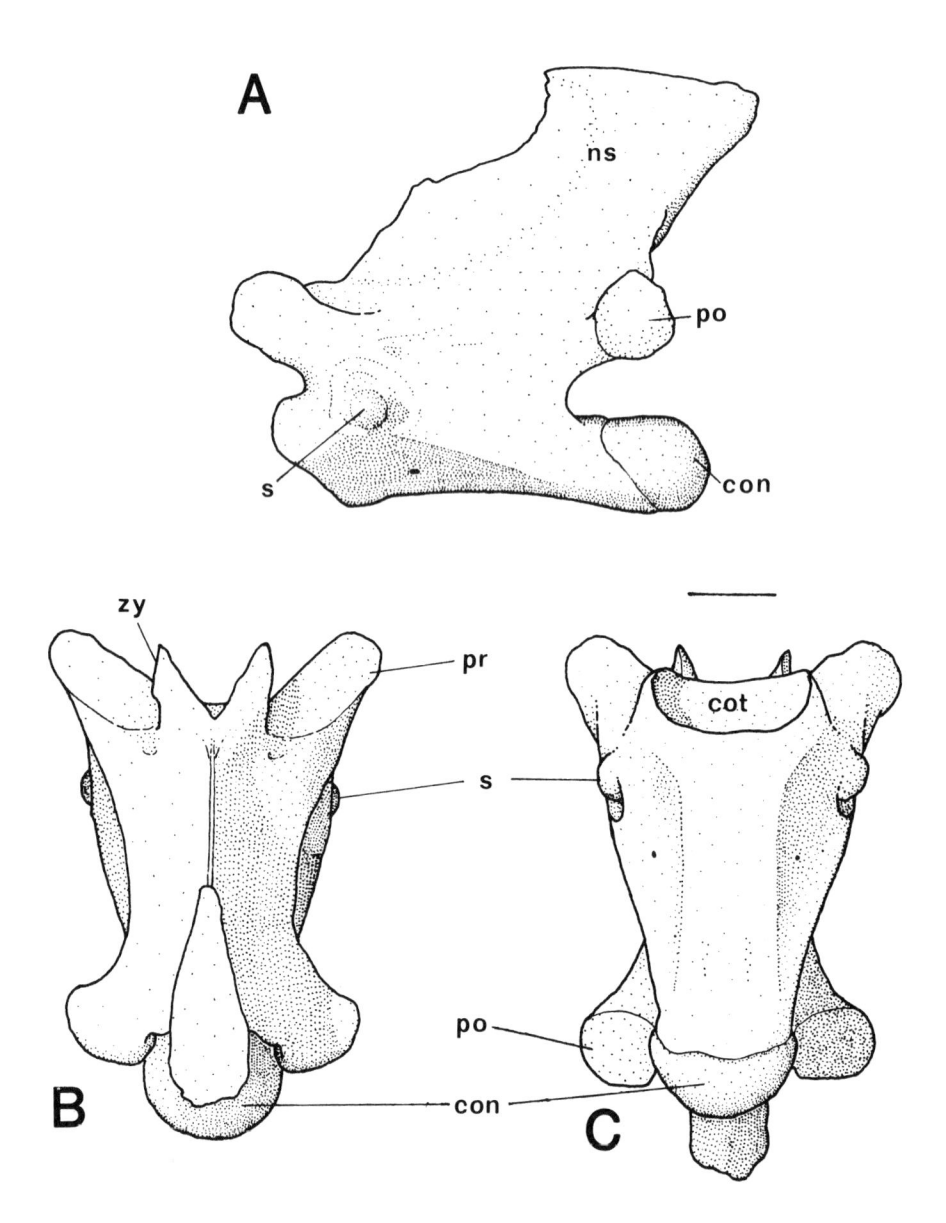

FIG. 34. Twentieth presacral vertebra of *Brachylophus vitiensis* (MCZ 160255) in (A) lateral (anterior to left), (B) dorsal, and (C) ventral views. Scale equals 2 mm. Abbreviations: con, condyle; cot, cotyle; ns, neural spine; po, postzygapophysis; pr, prezygapophysis; s, synapophysis for articulation of rib; zy, zygosphene.

TABLE 4. Numbers of Presacral Vertebrae

Taxon	N	23	24	25	26	27
Amblyrhynchus cristatus	11		*100%*			
Brachylophus fasciatus	13		*77*	23%		
vitiensis	4		*100*			
Conolophus pallidus	9		*100*			
subcristatus	11	9%	*91*			
Ctenosaura acanthura	5		*80*	20		
bakeri	12		*100*			
clarki	10			*90*	10%	
defensor	5			*40*	20% 20	20%
hemilopha	22		*100*			
palearis	8		*100*			
pectinata	13		*100*			
quinquecarinata	12		8	*92*		
similis	6		*83*	17		
Cyclura carinata	4	25	*75*			
collei	0	-	-	-	-	-
cornuta	13		*100*			
cychlura	3		*100*			
nubila	8		*100*			
pinguis	1		*100*			
ricordii	4		*100*			
rileyi	1		*100*			
Dipsosaurus dorsalis	35	3	*94*	3		
Iguana delicatissima	3		*100*			
iguana	20		*95*	5		
Sauromalus ater	5		*80*	20		
australis	2		50	*50%*		
hispidus	15		*93*	7		
obesus	23	4%	*83*	13		
slevini	3		*100*			
varius	10		*100*			

Note: The figures given are percentages of the total number of specimens. Modes are in italics. Numbers between columns represent specimens with sacral asymmetries.

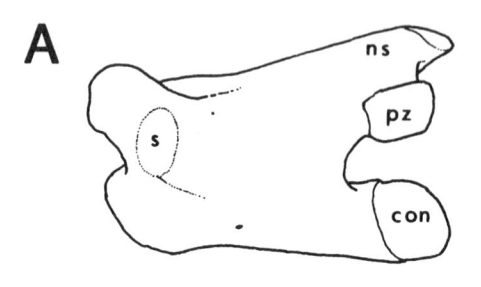

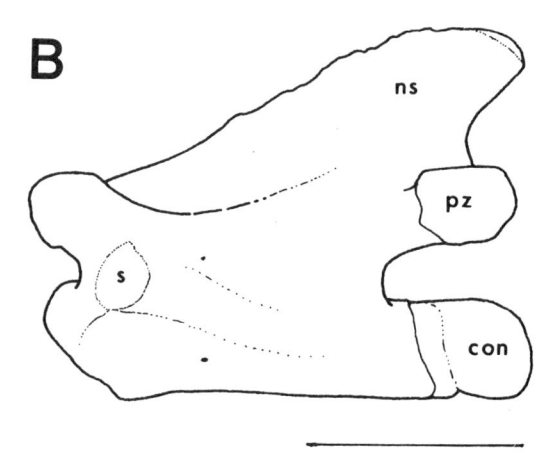

72   *University of California Publications in Zoology*

FIG. 35. Lateral views of the twentieth presacral vertebrae of (A) *Sauromalus obesus* (RE 1578) and (B) *Ctenosaura pectinata* (RE 641), showing differences in the height of the neural spine. Scale equals 0.5 cm. Abbreviations: con, condyle; ns, neural spine; pz, postzygapophysis; s, synapophysis.

Sauromalus differs from other iguanines in the morphology of its presacral vertebrae. In this genus, the neural spines of the presacral vertebrae are short (Fig. 35A); from the base of the postzygapophysial articular surfaces to the top of the neural spine they measure less than 50% of the total height of the vertebrae. In most other iguanines the neural spines make up more than 50% of the total vertebral height (Figs. 34A, 35B), though there is considerable variation in this category. This variation includes both interspecific differences in adult morphology and ontogenetic increase in neural spine height within species. *Ctenosaura* bridges the morphological gap between the two categories, with some members (e.g., *C. clarki*) approaching the condition seen in *Sauromalus*. Outgroup comparison yields equivocal results concerning the polarity of the different conditions of

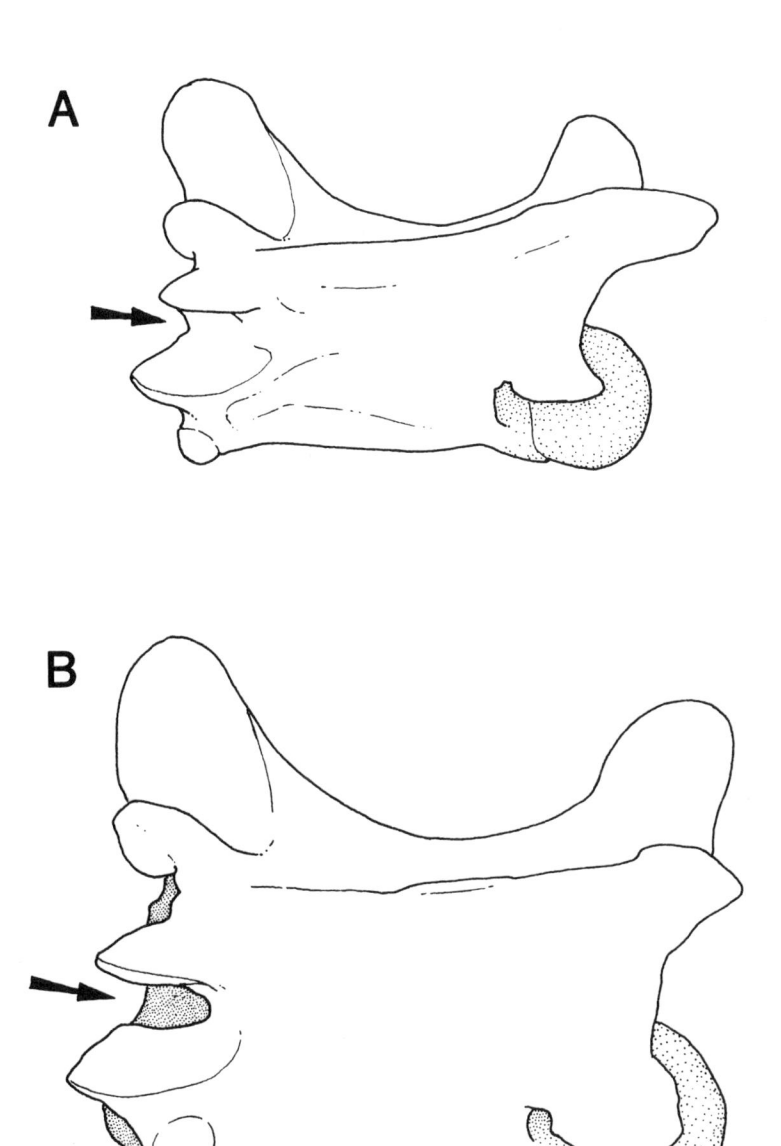

FIG. 36. Dorsolateral views of the twentieth presacral vertebrae of (A) *Dipsosaurus dorsalis* (KdQ 22) and (B) *Sauromalus obesus* (RE 1578), showing absence and presence, respectively, of bony separation (arrows) between the prezygapophyses and the zygosphenes. Scale equals 1 mm.

neural spine height. Crotaphytines, *Hoplocercus*, *Chalarodon*, and some *Oplurus* have short neural spines; those of *Laemanctus*, *Morunasaurus*, and other *Oplurus* are roughly intermediate; and those of *Basiliscus*, *Corytophanes*, and *Enyalioides* are tall, reaching extreme heights in adult male *Basiliscus*. Because of this ambiguous evidence, I did not use neural spine height as a character at the first level of phylogenetic analysis within iguanines, though it was used later at a lower hierarchical level.

The zygosphenes of *Dipsosaurus* differ from those of other iguanines (Fig. 36). In this taxon, the articular surfaces of the zygosphenes are connected laterally to those of the prezygapophyses by a continuous arc of bone (Fig. 36A). All other iguanines have a deep anterior notch separating the articular surfaces of the zygosphenes from those of the prezygapophyses (Fig. 36B).

In their weakest form, zygosphenes are mere out-turnings of the medial surfaces of the prezygapophysial facets that face dorsolaterally (Hoffstetter and Gasc, 1969). When more strongly developed, the articular surfaces of the zygosphenes are oriented laterally or ventrolaterally, eventually coming to face directly opposite those of the prezygapophyses. The final stage in the expression of the zygosphenal half of the accessory vertebral articulation appears to be the separation of the zygosphenes from the prezygapophyses by a notch. Thus, *Dipsosaurus* is the only iguanine that does not exhibit full development of the zygospheneal articulations. Although the degree to which the zygosphene-zygantrum articulation is developed may be positively correlated with size in iguanids (Etheridge, 1964a), this fact alone cannot account for its relatively weak development in *Dipsosaurus*, the smallest iguanine. Outside of Iguaninae, *Corytophanes*, which is about the same size (snout-vent length) as *Dipsosaurus*, possesses the deep notch separating zygosphenes from prezygapophyses, while *Petrosaurus* that are larger than *Dipsosaurus* do not.

Outgroup comparison provides equivocal evidence concerning the plesiomorphic zygosphenal morphology for iguanines. Among the outgroups examined in this study, the vertebrae of basiliscines and some *Enyalioides* resemble those of most iguanines in having strongly developed zygosphenes and zygantra with deep anterior notches between the articular surfaces of the zygosphenes and those of the prezygapophyses. Crotaphytines and most morunasaurs have weakly developed accessory vertebral articulations: the articular surfaces of the zygosphenes are continuous with the medial portions of those of the prezygapophyses, and, unlike those of all iguanines, they face dorsolaterally rather than ventrolaterally. The zygosphene-zygantrum articulations are very weakly developed in *Oplurus* and *Chalarodon*. Therefore, some nonhomology between morphologically similar vertebrae is required under the assumption of iguanine monophyly. Either the notch in the basiliscine accessory articulation (and that of some *Enyalioides*) is convergent with the one in iguanines, or its absence in *Dipsosaurus* is convergent (and possibly also a reversal) with a similar condition seen in other outgroups.

Sacrum (Fig. 39). Like all tetrapodous squamates, iguanines characteristically have two sacral vertebrae, although some specimens have asymmetrical sacra of the form reported by Hoffstetter and Gasc (1969) involving three vertebrae (Table 4). I recognize two characters in the sacra of iguanines, both involving the pleurapophyses of the posterior sacral vertebra.

The posterior edges of the pleurapophyses of the posterior sacral vertebrae of iguanines may or may not bear posterolaterally directed processes (Hofstetter and Gasc, 1969: Fig. 50). These processes are usually present, though not invariably so, in *Amblyrhynchus*, *Brachylophus*, *Conolophus*, *Dipsosaurus*, and *Sauromalus*, and are present in the single specimen of *Cyclura pinguis* examined; they are absent in *Ctenosaura*, *Iguana*, and other *Cyclura*. When present, each process lies posteroventral to a foramen in the posterior surface of the second pleurapophysis. Occasionally, a process may develop dorsolateral to the foramen; this process and the one described previously do not seem to be homologous on positional grounds.

Given the outgroups used in this study and their uncertain relationships, outgroup analysis is useless for assessing the plesiomorphic condition of this character. The processes are absent in basiliscines and *Hoplocercus*, present in the *Enyalioides*, variably present in *Oplurus*, *Chalarodon*, *Gambelia*, and *Morunasaurus*, and present in *Crotaphytus*. Therefore, I did not employ this character in phylogenetic analysis at the level of all iguanines.

The canal leading to the foramen that emerges alongside the posterior edge of each posterior sacral pleurapophysis has its medial opening on the ventral surface of the same pleurapophysis. This ventral foramen is almost always present in all iguanines except *Conolophus*. In *Conolophus*, the ventral foramen may also be present, but more often it is absent, and an open groove is left in place of the enclosed canal. The condition seen in *Conolophus* is almost certainly apomorphic, since all four outgroups generally possess the foramen and enclosed canal.

Caudal Vertebrae (Figs. 37, 38). Iguanine caudal vertebrae are highly variable, but possess many common structural features. The neural spines of the anterior caudal vertebrae are taller than their presacral counterparts, but they gradually decrease in size posteriad and increase their posterior orientation until they vanish toward the end of the tail. Complete haemal arches, positioned intercentrally, begin between the centra of the second and third or the third and fourth caudal vertebrae. They are oriented posteroventrally and, like the neural spines, decrease in size and increase in posterior orientation, moving posteriorly, until they vanish near the end of the tail. The bases of the haemal arches may form continuous basal bars or they may be separate. Small, paired elements, presumably serially homologous with the bases of the haemal arches, or otherwise incomplete haemal arches, often precede the first complete arch.

Four vertebral series (Fig. 37) can be recognized in the caudal sequence of iguanines (Etheridge, 1967). The anterior seven to fifteen caudal vertebrae bear a single pair of laterally or posterolaterally oriented transverse processes (fused caudal ribs) and lack autotomy septa (fracture planes) (Fig. 37A). In the following series, each vertebra bears two pairs of transverse processes that are either parallel or diverge from one another (Fig. 37B). The vertebrae in this second series and the remaining two series may or may not have autotomy septa. Species that lack autotomy septa generally have a shorter double-

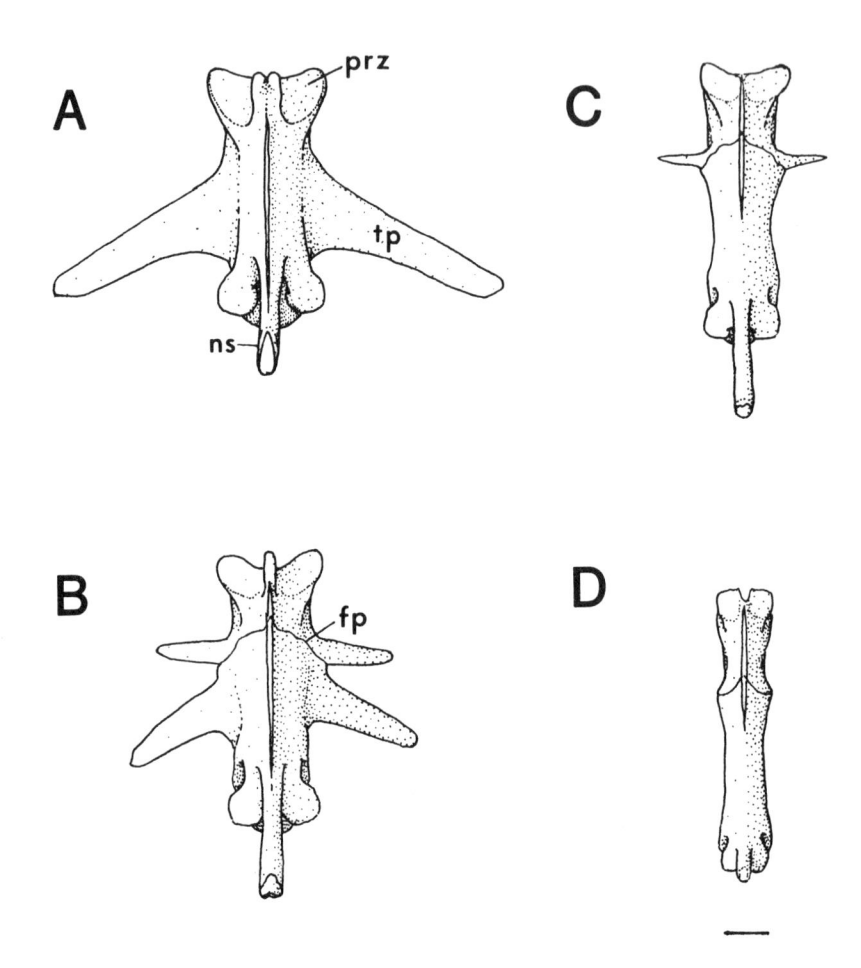

FIG. 37. Dorsal views of caudal vertebrae of *Dipsosaurus dorsalis* (KdQ 22): (A) number 4, (B) number 9, (C) number 15, and (D) number 28. Scale equals 1 mm. Abbreviations: fp, fracture plane; ns, neural spine; prz, prezygapophysis; tp, transverse process.

process series and more frequently possess bilaterally asymmetrical transverse processes. The transverse processes decrease in size posteriorly and, although the members of the posterior pair are as large or larger than those of the anterior pair, it is usually the former that disappear first (although the alternative is not uncommon), resulting in a third series with a single pair of transverse processes (Fig. 37C). These processes, presumably serially homologous with the anterior transverse processes of the second series, based on their anterior position on the vertebrae, continue to decrease in size until they vanish, leaving a fourth series whose vertebrae lack transverse processes (Fig. 37D). A variable number of vertebrae at the end of this last series are nonautotomic.

The number of caudal vertebrae in iguanines varies from as few as 25 in *Ctenosaura defensor* to over 70 in *Iguana iguana*. Because this number varies considerably within species, much of the variation is difficult to partition into character states nonarbitrarily. Nevertheless, an apparent gap exists between *Sauromalus* and some *Ctenosaura*, which have fewer than 40 caudal vertebrae, and all other iguanines, which have more than this number.

Outgroup comparison does not clearly indicate the plesiomorphic number of caudal vertebrae in iguanines. Most outgroup species have numbers of caudal vertebrae near or bridging the gap seen in iguanines. *Hoplocercus* is unique among outgroup taxa in having a very short (fewer than 20 vertebrae), spiny tail, even more extreme than those of certain *Ctenosaura*, and lacking any complete haemal arches. Because of this ambiguity, I used the number of caudal vertebrae as a systematic character only at a level less inclusive than all iguanines.

Unlike other iguanines, *Amblyrhynchus*, *Brachylophus*, *Conolophus*, and *Iguana delicatissima* lack autotomy septa along their entire caudal sequences throughout postembryonic ontogeny, and thus presumably are unable to autotomize their tails. This does not mean, however, that these lizards cannot regenerate their tails, for caudal regeneration occurs in both *Brachylophus fasciatus* (Etheridge, 1967) and *B. vitiensis*. In these cases, regeneration was associated with a broken vertebra rather than intervertebral separation, supporting Etheridge's (1967) suggestion that regeneration is a function of trauma to the vertebra rather than autotomy itself (but see Bellairs and Bryant, 1985). It is noteworthy that all iguanines that lack caudal fracture planes are insular forms. Caudal autotomy is generally thought to be an adaptation for escaping predators (Congdon et al., 1974; Turner et al., 1982), and the intensity of predation is often less severe on islands (Carlquist, 1974).

I am unable to resolve the polarity of this character with the four outgroups used in this study. The basiliscines *Laemanctus* and *Corytophanes*, the crotaphytine *Crotaphytus*, and the morunasaur *Hoplocercus* lack autotomy septa, but in other members of all of these groups and in all oplurines examined, the septa are present. Thus, monophyly of each of the outgroups and of iguanines requires multiple homoplastic events no matter which condition, presence or absence of autotomy septa, is considered to be plesiomorphic for iguanines. Because of the ambiguity involved in this character, I withheld an initial decision on its polarity and used it only at a hierarchical level below that of all iguanines.

The beginning of the second series of caudal vertebrae varies both within and among iguanine species. High overlap among species in the range of this character within species renders much of this variation useless as systematic characters, but one character can be recognized for the purpose of comparisons among the basic taxa used in this study. In *Brachylophus* and *Dipsosaurus*, the series of caudal vertebrae with two pairs of transverse processes per vertebra begins at the eighth to the tenth caudal vertebra; in all other iguanines, this series begins at the tenth or a more posterior vertebra. Because of intraspecific variation in the beginning of this second series of caudal vertebrae, a given specimen may not be assignable to one or the other group, but a species (sample) can be so assigned.

Unfortunately, the pathway of character-state transformation cannot be analyzed by outgroup comparison without making additional assumptions about the character. None of the four outgroups used in this study, nor any other iguanian, possesses caudal vertebrae with two pairs of transverse processes (Etheridge, 1967). Nevertheless, a close correspondence between the beginning of the series of caudal vertebrae with two pairs of transverse processes and the beginning of the series of autotomic vertebrae in iguanines suggests that the latter might be used as the character instead. Unfortunately, not all iguanines (nor all outgroup taxa) possess autotomic caudal vertebrae. Therefore, in order to use this character I first must assume that the beginning of the series of caudal vertebrae with two pairs of transverse processes in taxa that lack autotomy septa corresponds with the beginning of the autotomic series in those taxa that possess autotomy septa. Second, I must assume that the beginning of the autotomic series in taxa that lack vertebrae with two pairs of transverse processes corresponds with the beginning of the series of vertebrae with two pairs of transverse processes.

Under these assumptions, outgroup comparison can be used with those outgroups possessing autotomic vertebrae, but it provides ambiguous evidence concerning the plesiomorphic condition of this character. The autotomic series of *Basiliscus* begins in a range that has the tenth caudal vertebra in its midst. That of *Gambelia* begins posterior to the tenth vertebra, while those of *Enyalioides*, *Morunasaurus*, and oplurines begin anterior to the tenth vertebra. The polarity decision for this character will thus vary depending upon the relationships among iguanines and the four outgroups. Because these relationships are unknown, I withheld a decision on the polarity of this character in phylogenentic analysis at the level of all iguanines.

Lazell (1973:1-2) citing Etheridge (in litt.) distinguished *Iguana* from *Cyclura* by the presence of "a low finlike process above the neural arch of no more than six anterior caudal vertebrae" in the former, compared to the "high, finlike processes above the neural arches of all the caudal vertebrae" in the latter. The processes in question are presumably ossifications of the dorsal skeletogenous septum. When the remaining iguanine genera are considered, there appears to be a continuum in the height of these processes rather than two discrete morphologies, low and high. Even within an organism, the morphology of these processes differs among the caudal segments. In most iguanines, the processes on the anterior caudal vertebrae are merely thin, midsagittal extensions of the anterior edges of the neural spines. Moving posteriorly along the column, apices form on the processes, and the processes themselves are displaced anteriorly, sometimes becoming entirely separated from their respective neural spines. The height of the processes increases, then gradually decreases, moving anterior to posterior. Although the midsagittal processes generally disappear short of the end of the tail, they are present (Fig. 38A) well beyond the anterior third of the caudal sequence (determined by vertebra number, not by distance from the beginning of the tail) in all genera except *Brachylophus* and *Iguana*. The situation in *Brachylophus* and *Iguana* differs from the one described above in that the processes are

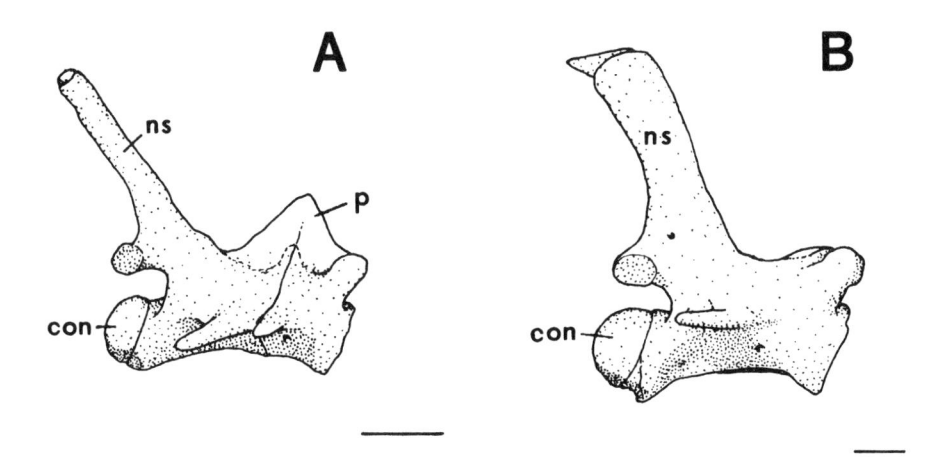

FIG. 38. Lateral views of the ninth caudal vertebrae of (A) *Dipsosaurus dorsalis* (KdQ 22) and (B) *Iguana iguana* (MVZ 78384), showing differences in the size of the dorsal midsagittal processes. Scale equals 2 mm; anterior is to the right. Abbreviations: con, articular condyle; ns, neural spine; p, dorsal midsagittal process.

relatively small and do not continue as far posteriorly in the caudal sequence (Fig. 38B). Although they may be present beyond the sixth caudal vertebra, I have never observed them beyond the tenth. The caudal sequences of *Brachylophus* and *Iguana* consist of more than 55 vertebrae; thus, the processes are not present beyond the anterior fifth of the sequence.

Although the evidence is somewhat equivocal, outgroup comparison favors the interpretation that the condition of the midsagittal processes of the caudal vertebrae seen in *Brachylophus* and *Iguana* is apomorphic. The alternative condition occurs in crotaphytines, morunasaurs, and oplurines, but basiliscines are similar to *Brachylophus* and *Iguana*. In basiliscines, the small, finlike processes are rarely found posterior to the fifth caudal vertebra. Basiliscines, *Brachylophus*, and *Iguana* are all arboreal, suggesting a possible functional relationship between the morphology of the caudal vertebrae and use of the tail in arboreality.

Ribs (Fig. 39). Variation in the numbers and the morphology of various kinds of ribs has served as the basis for characters in previous systematic studies of iguanids (Etheridge, 1959, 1964a, 1965b, 1966); but iguanines are conservative in most of these features. Like those of all iguanids, iguanine ribs are holocephalous and most have two parts: a bony dorsal portion and a cartilaginous ventral portion, the inscriptional rib (Etheridge, 1965b). The length of the inscriptional ribs is highly variable from one region of the vertebral column to another, and at the posterior end of the presacral series these elements are often lacking.

Cervical ribs, those ribs anterior to the first ribs that are attached to the sternum, typically number four pairs in iguanines, beginning on the fifth presacral vertebra (very

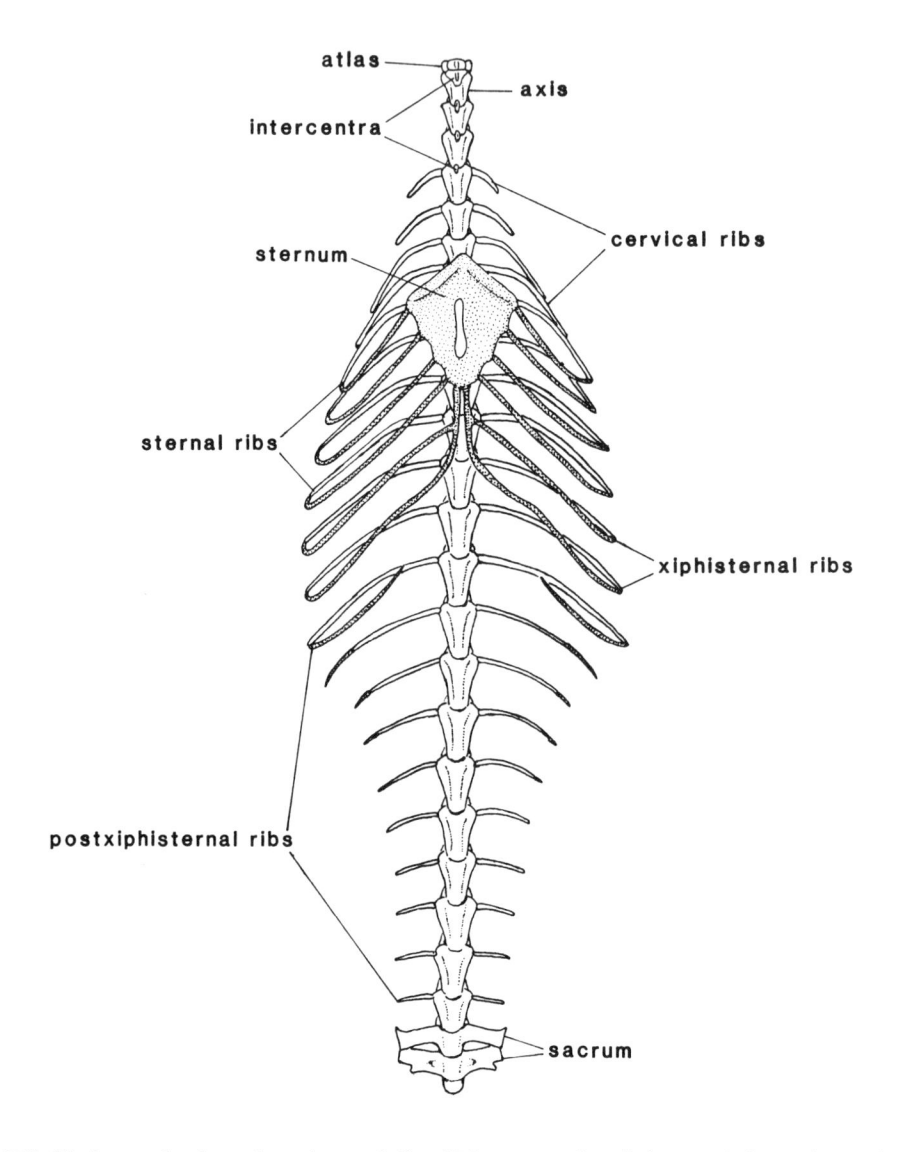

FIG. 39. Presacral and sacral vertebrae and ribs of *Dipsosaurus dorsalis* in ventral view. The drawing is a composite.

rarely on the fourth) and ending on the eighth. The bony portions of the first two cervical rib pairs are short, while the second two are much longer, about the same length as the anterior thoracic ribs. The next four (rarely three) rib pairs, on presacral vertebrae nine through twelve, are sternal ribs, attached ventromedially to the lateral borders of the sternum through their cartilaginous ventral portions. Two (rarely three; sometimes one in *Sauromalus*) pairs of xiphisternal ribs follow the sternal ribs. These ribs articulate dorsally with vertebrae 13 and 14, and their cartilaginous ventral portions unite with one another before attaching to the posterior end of the sternum. The remaining ribs are simply termed postxiphisternal. The bony anterior postxiphisternal ribs are often as long as their xiphisternal counterparts, but there is a progressive reduction in their length posteriorly. The posteriormost ribs are shorter than the sacral pleurapophyses. Lumbar vertebrae, posterior presacral vertebrae lacking ribs, are not found in iguanines. Very rarely, the ribs of the posteriormost presacral segment are fused to the vertebra.

Etheridge (1965b) described variation in the abdominal skeleton (postxiphisternal inscriptional ribs) of iguanids. All iguanines were reported to exhibit a pattern in which all postxiphisternal inscriptional ribs are attached to their corresponding dorsal bony ribs. In some iguanines, all of these inscriptional ribs end free, while in others the members of one or more of the anterior pairs may join midventrally to form continuous chevrons. Based on Etheridge's (1965b) findings and my own observations, the iguanine genera exhibit the following morphologies in the abdominal skeleton: (1) continuous chevrons absent (*Dipsosaurus*, *Sauromalus*); (2) continuous chevrons present or absent (*Amblyrhynchus*, *Conolophus*, *Ctenosaura*, *Cyclura*, *Iguana*); and (3) continuous chevrons present (*Brachylophus*). The number of continuous chevrons and other enlarged postxiphisternal inscriptional ribs may exhibit taxon-specific patterns, but because the fragile abdominal skeleton is often destroyed in skeletal preparations, I have not been able to examine enough specimens to assess these patterns adequately.

In the outgroups that I have examined, postxiphisternal inscriptional ribs that form continuous midventral chevrons are found only in morunasaurs; however, because they share the common feature of having at least some inscriptional ribs that bear no traces of attachment to the bony ribs, Etheridge (pers. comm.) believes that the oplurine pattern is a transformation of that seen in morunasaurs. Basiliscines and crotaphytines are similar to *Dipsosaurus* and *Sauromalus* in their lack of continuous chevrons. Thus, evidence bearing on the polarity of this character is equivocal, and I did not use it in my initial analysis of relationships among iguanine genera.

PECTORAL GIRDLE AND STERNAL ELEMENTS

The iguanine pectoral girdle and sternal elements (Fig. 40) are closely associated and form a complex functional unit composed of six pairs of elements plus two median, unpaired ones. Some of these elements are composed entirely of calcified cartilage, while others are bony. All iguanines possess all 14 elements: suprascapulae, scapulae, coracoids, epicoracoids, clavicles, interclavicle, sternum, and xiphisterna.

Suprascapulae (Fig. 40). These are paired fan-shaped elements composed of calcified cartilage that extend continuously from the dorsal edges of the scapulae. The suprascapulae lie just external to the posterior cervical and the anterior thoracic bony ribs. They are not attached directly to the axial skeleton, but ride over the bony portions of the ribs. As in most squamates, the only direct skeletal attachments between pectoral girdle and axial skeleton are through the sternum and cartilaginous portions of the anterior thoracic ribs. In

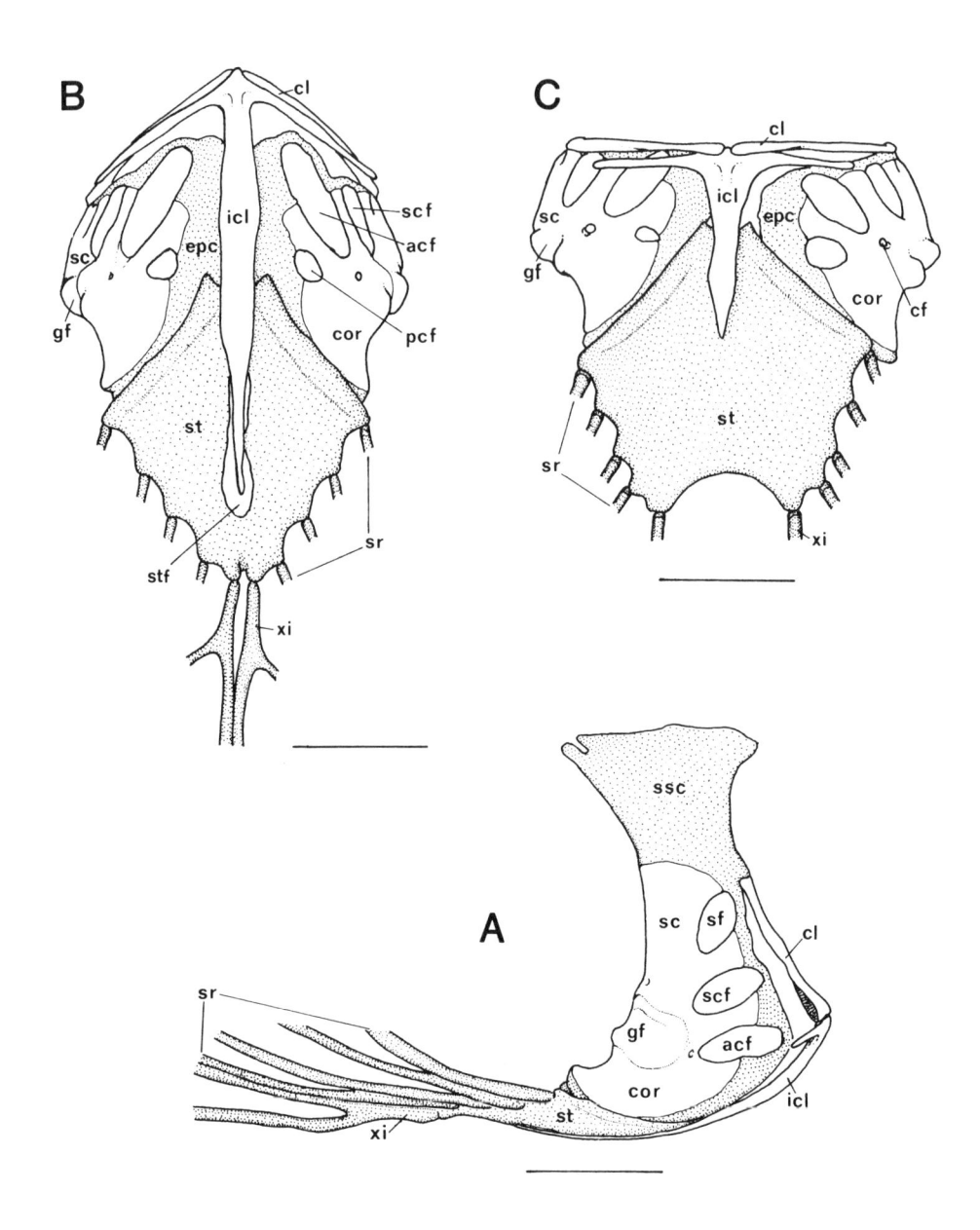

FIG. 40. Pectoral girdles of (A) *Brachylophus fasciatus* (RE 1866), (B) *Ctenosaura hemilopha* (RE 1341), and (C) *Sauromalus obesus* (RE 411). A is a lateral view; anterior is to the right. B and C are ventral views. Calcified cartilage is stippled. Scale equals 1 cm. Abbreviations: acf, anterior coracoid fenestra; cf, coracoid foramen; cl, clavicle; cor, coracoid; epc, epicoracoid; gf, glenoid fossa; icl, interclavicle; pcf, posterior coracoid fenestra; sc, scapula; scf, scapulocoracoid fenestra; sf, scapular fenestra; sr, sternal ribs; ssc, suprascapula; st, sternum; stf, sternal fontanelle; xi, xiphisternum.

most iguanines, the surfaces of the scapulae and suprascapulae form a continuous, laterally convex arc, but in *Sauromalus* the junction of these surfaces is angular and the suprascapulae are oriented more horizontally than in other iguanines. The condition of the suprascapulae in *Sauromalus* is presumably related to the depressed body form of these animals, and on the basis of outgroup comparison is almost certainly apomorphic.

Scapulae, Coracoids, and Epicoracoids (Fig. 40). The scapula and coracoid of each side are closely associated and function as a single unit. Although separated by a suture throughout most of the period of growth, the two bones fuse to form a single scapulacoracoid element near the attainment of maximum size. Prominent features of the scapulocoracoids are the glenoid fossae for the articulation of the humeri, which lie at the junctions between scapulae and coracoids along their posterior edges, coracoid foramina anteroventral to the glenoid fossae, and three or four (rarely two) scapulocoracoid fenestrations on each side of the girdle, the functional significance of which is discussed by Peterson (1973).

The scapulocoracoid fenestrations pierce the pectoral girdle along the anterior margins of the scapulae and coracoids, between these bones and the cartilagenous epicoracoids (Fig. 40). Following the terminology of Lécuru (1968a), from dorsal to ventral the four pairs of fenestrations are: (1) scapular fenestrae, which lie anterodorsally within the scapulae; (2) scapulocoracoid fenestrae, situated at the junctions between scapulae and coracoids; (3) anterior (primary) coracoid fenestrae, located within the coracoids; and (4) posterior (secondary) coracoid fenestrae, also located within the coracoids but posteroventral to the anterior coracoid fenestrae. All iguanines invariably possess the scapulocoracoid and the anterior coracoid fenestrae; the scapular fenestrae and the posterior coracoid fenestrae may be present or absent.

Scapular fenestrae are invariably present in all iguanines except *Amblyrhynchus* and *Sauromalus*, in which they are small or occasionally absent. Outgroup analysis yields equivocal results concerning the polarity of these character states. Scapular fenestrae are present in crotaphytines, the single *Enyalioides oshaughnessyi* examined, *Chalarodon*, and *Oplurus cuvieri*; they are absent in basiliscines, other morunasaurs, and *Oplurus quadrimaculatus* (in which the large "scapulocoracoid" fenestrae may be homologous with the scapular plus the scapulocoracoid fenestrae of other oplurines). Because of this ambigiuty, I used the presence or absence of scapular fenestrae as a systematic character only at a level less inclusive than all iguanines.

The presence of posterior coracoid fenestrae is more variable intragenerically than the presence of scapular fenestrae. Posterior coracoid fenestrae are invariably absent in *Brachylophus* (Fig. 40A); usually absent in *Dipsosaurus*; usually present in *Amblyrhynchus*, *Ctenosaura* (Fig. 40B), *Cyclura*, and *Sauromalus* (Fig. 40C); and invariably present in *Conolophus* and *Iguana*. The amount of variability differs among the genera in the third group. Posterior coracoid fenestrae are frequently absent in *Amblyrhynchus* and *Sauromalus,* in which all species are variable in the presence of these fenestrae except *S. australis* and *S. slevini*, both of which are represented by small samples (n=2). The absence of a posterior coracoid fenestra is rare in *Ctenosaura*; it has been detected in only some members of three species, *C. clarki*, *C. hemilopha*, and *C. similis*. In *Cyclura*, the absence of a posterior coracoid fenestra was observed only in two out of eight *C. nubila*, one of which lacked the fenestra unilaterally.

According to Peterson (1973), the presence of a posterior coracoid fenestra is associated with large size and/or the presence of a proximal belly of the *M. biceps*. Because a posterior coracoid fenestra is present in the species of *Ctenosaura* that reach smaller maximum sizes than *Brachylophus*, in which the fenestra is absent, presence of the fenestra cannot be strictly size-dependent. The association of the fenestra with a proximal belly of the *M. biceps* was not examined in the present study.

Although the evidence is somewhat ambiguous, outgroup comparison favors the interpretation that the absence of posterior coracoid fenestrae is plesiomorphic for iguanines. Basiliscines and oplurines invariably lack these fenestrae. Morunasaurs generally lack posterior coracoid fenestrae, but in rare cases very small ones are present. Crotaphytines generally possess posterior coracoid fenestrae, although they are occasionally absent in *Gambelia*. If the general rather than the invariable presence or absence of posterior coracoid fenestrae is considered to be the systematic character, then outgroup comparison will either yield equivocal results or indicate that the absence of posterior coracoid fenestrae is plesiomorphic, depending on the relationships among iguanines and the four outgroups.

Clavicles (Fig. 40). Iguanine clavicles are boomerang-shaped, paired bones lying along the anterior margin of the pectoral girdle. They articulate ventromedially with the anterior median end of the interclavicle and dorsolaterally with the anteroventral edges of the suprascapulae. Compared with those of certain other iguanids, the clavicles of iguanines are relatively simple, generally lacking sharp, ventrally directed processes (hooks of Etheridge, 1964a) and ventromedial fenestrae, although small fenestrae are sometimes present in *Conolophus*.

Sauromalus differs from other iguanines in having slender clavicles, which are more or less elliptical in cross section. The clavicles of other iguanines have thin lateral shelves, making them wider when viewed anteriorly, although some *Ctenosaura* approach the condition seen in *Sauromalus*. Because the clavicles of all outgroup taxa examined except *Oplurus quadrimaculatus* are wide with thin lateral shelves, this condition must be considered plesiomorphic for iguanines.

Interclavicle (Fig. 40). This median, unpaired bone is the ventralmost in the pectoral girdle. In iguanines it bears the shape of a "T" or an arrow, formed by a lateral process at the anterior end on each side and a median posterior process. The anterior process seen in certain other squamates (Lécuru, 1968b) is virtually absent.

The extent of the posterior median process of the interclavicle varies among iguanines and is here assessed by the location of the posterior tip of the bone relative to the lateral corners of the sternum and the sternal attachments of the cartilaginous sternal ribs. *Amblyrhynchus* and *Sauromalus* (Fig. 40C) have short interclavicles that do not extend posteriorly beyond the lateral corners of the sternum, where the first pair of sternal ribs attaches. In all other iguanines except *Conolophus pallidus* and *Cyclura nubila* the posterior process of the interclavicle extends beyond this level (Fig. 40B) and, depending on the taxon, it may extend beyond the points of attachment of the second or even the third sternal-rib pairs. *Conolophus pallidus* and *Cyclura nubila* have interclavicles of intermediate length. In these taxa the interclavicle extends to about the level of the lateral corners of the sternum or slightly beyond. The width of the posterior process appears to be related to its posterior extent: short interclavicles are usually wider than long ones. The correlation is not strict, however, for some *Sauromalus* have narrow posterior processes.

Among the outgroups examined, only some *Crotaphytus* have an interclavicle that does not extend posteriorly beyond the lateral corners of the sternum. I therefore considered the short interclavicle to be apomorphic for iguanines.

Another variable feature of iguanine interclavicles is the angle between each lateral process and the posterior process. All species exhibit at least 10° of variation in this feature with significant intertaxic overlap. For this reason I recognize only two categories as character states. *Amblyrhynchus* and *Sauromalus* (Fig. 40C) have roughly T-shaped interclavicles, with the angle between the lateral and posterior processes ranging from 75° to 90°. Other iguanines have arrow-shaped interclavicles (Fig. 40B); the angle formed by the lateral and posterior processes is usually less than 75°. Although the angle in question overlaps the first category in some members of both species of *Brachylophus* and *Conolophus,* as well as in some *Cyclura nubila*, the lower limits of the range of angles in these species is well below that in *Amblyrhynchus* and *Sauromalus*. Outgroup comparison indicates that the arrow-shaped interclavicle is plesiomorphic. Among basiliscines, crotaphytines, morunasaurs, and oplurines, I have found T-shaped interclavicles only in the basiliscines *Laemanctus serratus* and *Corytophanes hernandesii*.

Sternum and Xiphisterna (Figs. 37, 40). The sternum of iguanines is shaped like a diamond or a pentagon and is composed of calcified cartilage. In embryos and some hatchlings, the sternal plate is paired, but the two halves fuse in late embryonic or early postembryonic ontogeny to form a single median element. Anterolaterally, the sternum meets the epicoracoids in a tongue-in-groove articulation, the coracosternal joint, which permits posterolateral-anteromedial movements of the scapulocoracoid units relative to the sternum (Jenkins and Goslow, 1983). The posterolateral borders of the sternal plate are the attachment sites for the cartilaginous ventral portions of four thoracic rib pairs (sternal ribs) and two others that attach via the xiphisterna. A sternal fontanelle may be present (Fig. 40B) or absent (Fig. 40C).

In most iguanines, the sternal fontanelle is long and narrow and is covered partially or completely by the posterior process of the interclavicle. In *Amblyrhynchus* and *Sauromalus* the sternal fontanelle is often small, and in the latter it may be subdivided into two or three small, round holes. In some specimens of both taxa the fontanelle is absent. Absence or small size of the sternal fontanelle is unequivocally apomorphic on the basis of the outgroups used in this study.

Sternal shape is variable in iguanines and is partly related to another feature, the proximity of the two sternal-xiphisternal attachments to one another and the midline. In most iguanines the xiphisterna attach to the sternum very close to the midline and to one

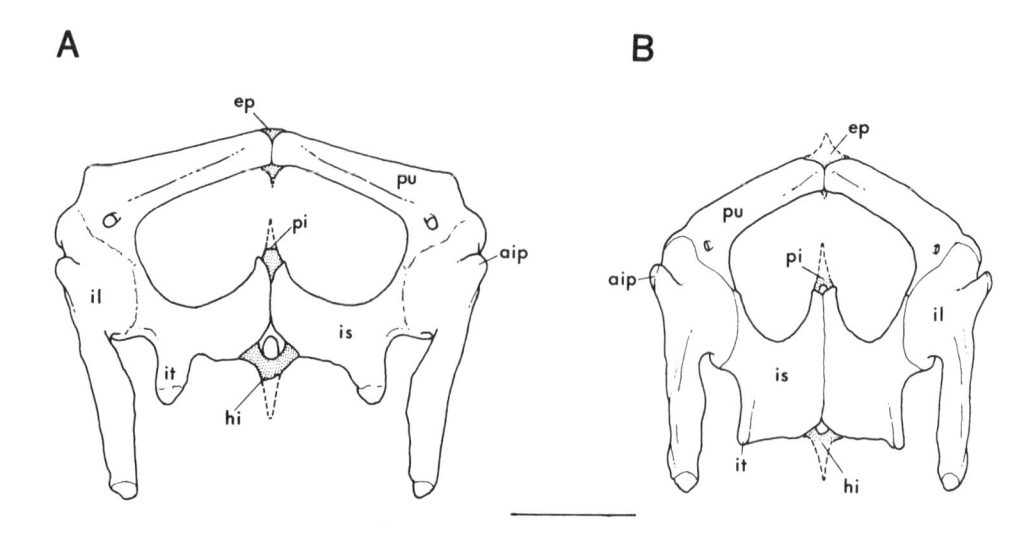

FIG. 41. Pelvic girdles of (A) *Sauromalus obesus* (RE 467) and (B) *Ctenosaura pectinata* (RE 419) in dorsal view. Scale equals 1 cm. Abbreviations: aip, anterior iliac process; ep, epipubis; hi, hypoischiac cartilage; il, ilium; is, ischium; it, ischial tuberosity; pi, proischiac cartilage; pu, pubis.

another, yielding a diamond-shaped sternum (Fig. 40B). In *Sauromalus* the xiphisterna are widely separated from one another, and the sternum is pentagonal (Fig. 40C). *Amblyrhynchus* is somewhat intermediate, having a small but distinct gap between its xiphisterna; however, the shape of its sternum is much closer to that of most other iguanines than to that of *Sauromalus*.

Most members of all outgroup taxa examined have diamond-shaped sterna with the xiphisterna in close proximity to each other. The exceptions are *Oplurus quadrimaculatus* and *Crotaphytus*, which approach the condition seen in *Sauromalus* to a greater or lesser degree, respectively. Although the pentagonal sternum with widely separated xiphisterna is probably apomorphic, the ambiguity is sufficient to force me to use this character only at a less inclusive level than that of all iguanines.

PELVIC GIRDLE

The iguanine pelvic girdle (Fig. 41) consists of three pairs of bones: dorsal ilia, which articulate with the sacral pleurapophyses; posteroventral ischia; and anteroventral pubes. Cartilaginous epipubes, and proischiac and hypoischiac cartilages, are situated on the midline between the pubes and the anterior and posterior parts of the ischia, respectively. An obvious difference in the shape of the pelvic girdle separates *Sauromalus* (Fig. 41A) from all other iguanines (Fig. 41B). Relative to those of other iguanines, the pelvis of *Sauromalus* is short and broad, clearly an apomorphic condition on the basis of the outgroups examined.

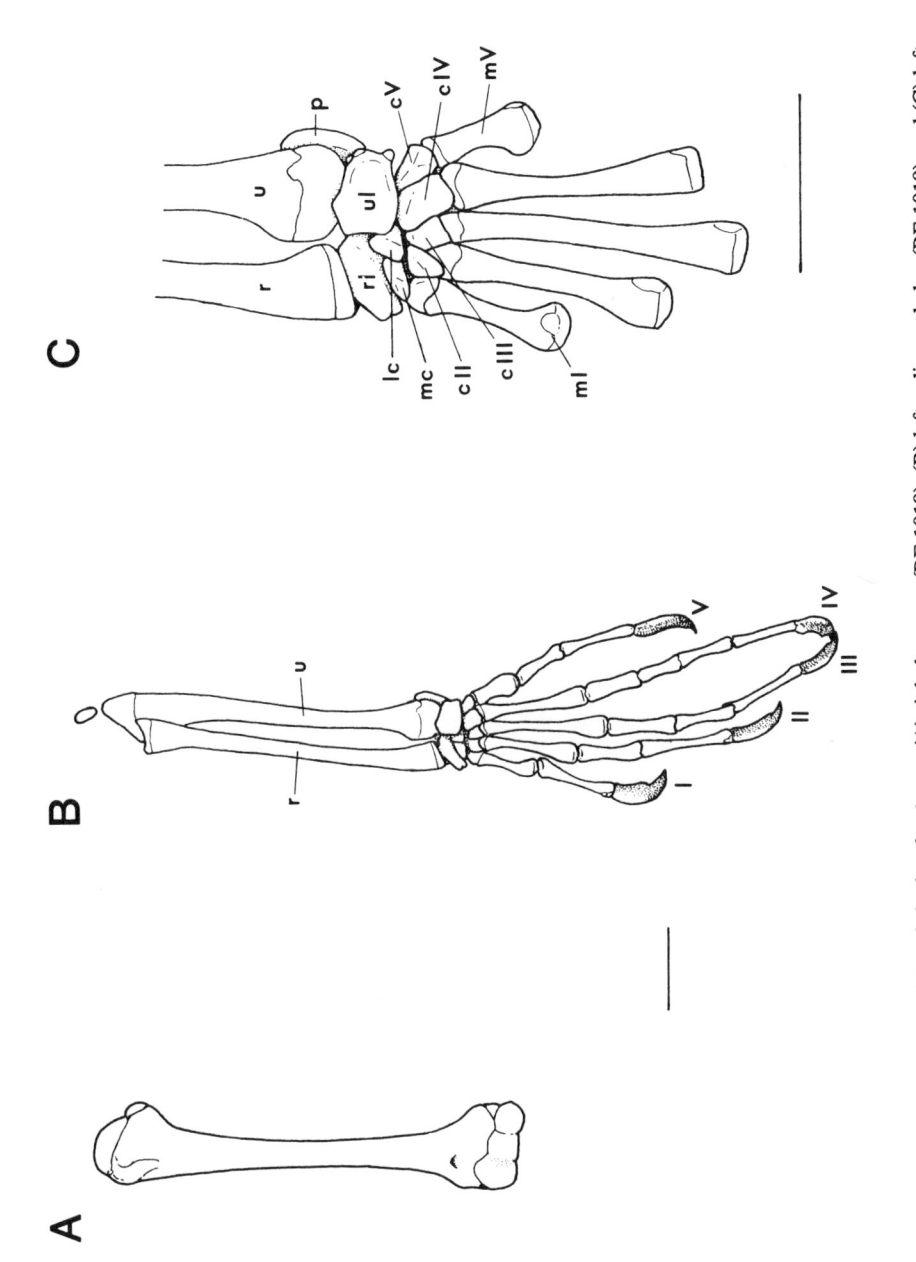

FIG. 42. Bones of the anterior limb of *Brachylophus fasciatus*: (A) right humerus (RE 1019), (B) left radius and ulna (RE 1019), and (C) left carpal region (RE 1866). Scale equals 0.5 cm. Abbreviations: cII-V, distal carpals 2-5; lc, lateral centrale; mc, medial centrale; m1 and m5, metacarpals 1 and 5; p, pisiform; r, radius; rl, radiale; u, ulna; ul, ulnare; I-V, digits 1-5.

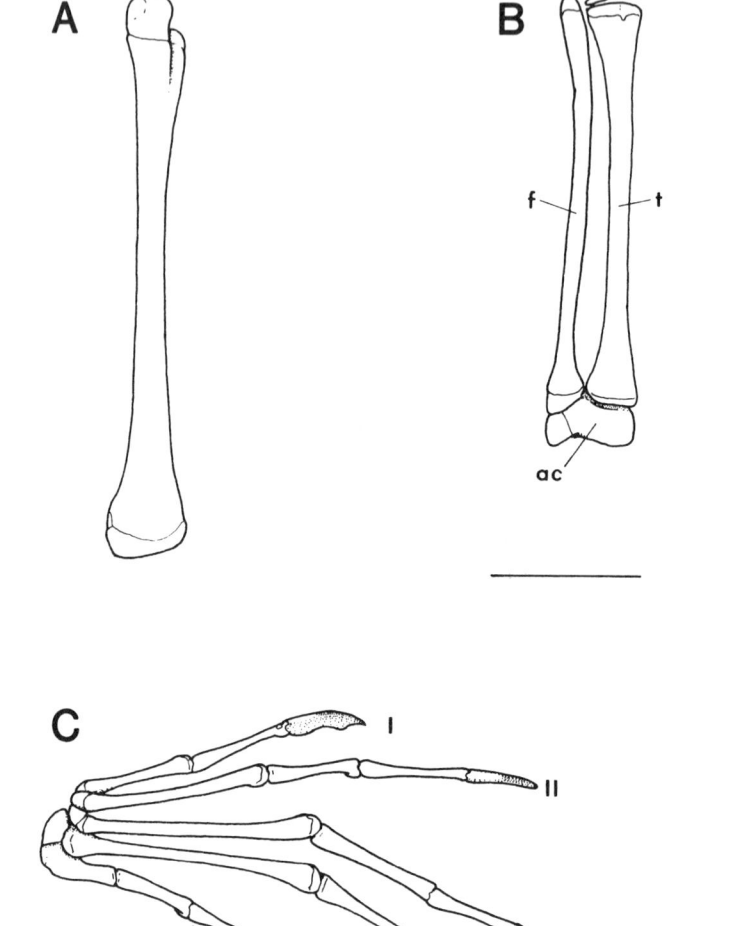

FIG. 43. Right hind limb skeleton of *Brachylophus fasciatus*: (A) femur; (B) tibia, fibula, and proximal tarsals; and (C) distal tarsals, metatarsals, and phalanges. Scale equals 1 cm. Abbreviations: ac, astragalocalcaneum; f, fibula; t, tibia; I-V, digits 1-5.

Another unique feature occurs in some *Sauromalus*, notably *S. varius*. In these animals the ischium is excavated mesial to the posteriorly directed ischiac tubercle, enhancing the distinctness of this structure. Because this character varies within a single genus, it is uninformative about relationships among the basic taxa used in this study.

I disagree with Lazell's (1973:1-2) statement that "In *Dipsosaurus* and *Sauromalus* the ilial shaft tapers abruptly posteriorly and the anterior iliac process is rather weakly developed." The ilial shaft of *Sauromalus* is narrower at its posterior terminus than those of other iguanines, but it does not taper abruptly. In *Dipsosaurus* the ilial shaft may taper abruptly, but it is broad near its posterior end like that of other iguanines except *Sauromalus*. While the anterior iliac process of *Sauromalus* does appear to be relatively small, that of *Dipsosaurus* is not.

LIMBS

Iguanine limbs exhibit considerable variation, but I have chosen not to use this variation as the basis for systematic characters. All iguanines possess the same bony elements in their limbs, but the proportions of the various limb bones vary considerably among iguanine taxa. Nevertheless, these proportions seem to be very plastic features, so plastic that I was unable to establish polarities with any confidence. Therefore, I give only a general description of this variation and devote most of the section to the description of characters that do not vary among iguanines but that may be useful at higher levels of comparison.

Compared to those of other iguanines, the limb bones of *Brachylophus* are relatively long, while those of *Amblyrhynchus* and *Sauromalus* are relatively short. These proportional differences are most evident in the long bones, metapodials, and phalanges. Proportional differences in the carpal and tarsal elements (mesopodials) are less obvious.

All iguanines possess the following bones in the forelimb (Fig. 42): humerus, radius, ulna, radiale, ulnare, pisiform, lateral centrale, five distal carpals, five metacarpals, and 17 phalanges. According to Carroll (1977), the first distal carpal of modern lizards is homologous with the medial centrale of other diapsids. As in other iguanids (Renous-Lécuru, 1973), the intermedium is absent. The phalangeal formula of the manus is 2:3:4:5:3. An entepicondylar foramen is present in the humerus.

The hind limbs of iguanines (Figs. 43, 44) consist of femur, tibia, fibula, astragalocalcaneum, two distal tarsals proximal to metatarsals three and four, five metatarsals, and 18 phalanges. The phalangeal formula of the pes is 2:3:4:5:4 which, like that of the manus, is presumably plesiomorphic for squamates.

OSTEODERMS

Two large *Amblyrhynchus* (JMS 126, 127) have dermal ossifications that apparently formed within the large, conical scales overlying the nasal, prefrontal, and frontal bones (Pl. 1), confirming Camp's (1923:307) observation that osteoderms are present in this taxon. Osteoderms, which differ from the rugosities that develop on various bones of the

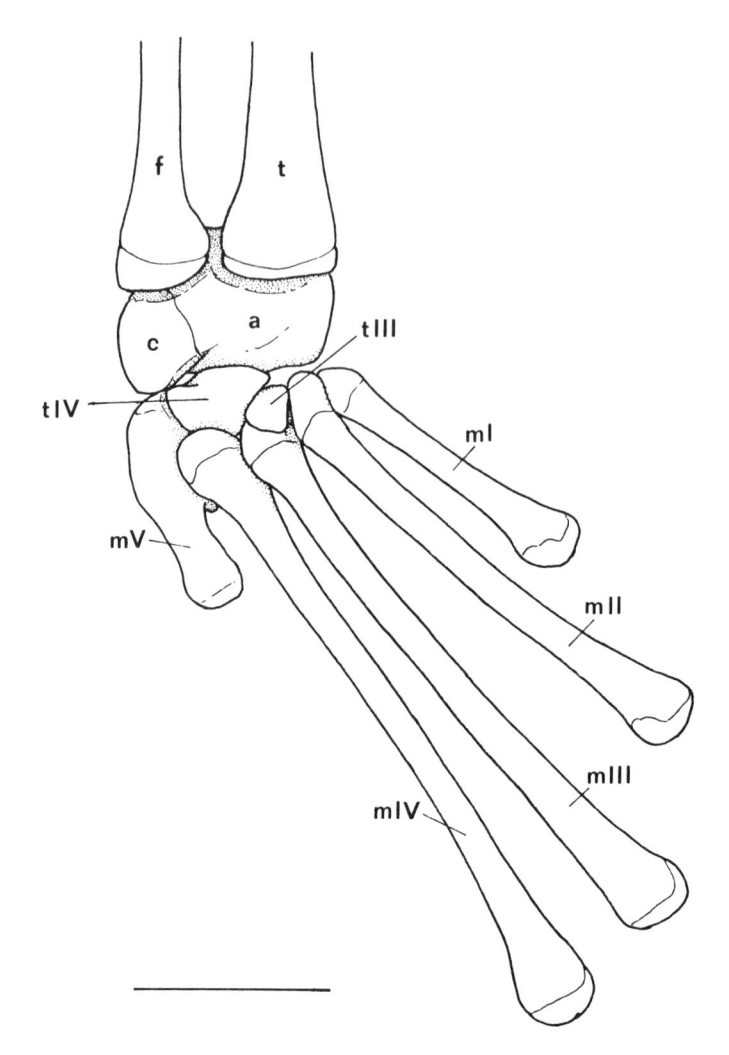

FIG. 44. Right tarsal region of *Brachylophus fasciatus*. Scale equals 0.5 cm. Abbreviations: a, astragalus; c, calcaneum; f, fibula; mI-V, metatarsals 1-5; t, tibia; tIII and tIV, distal tarsals 3 and 4.

dermal skull roof in certain iguanids, are unknown in iguanids other than *Amblyrhynchus* (Etheridge and de Queiroz, 1988), and their presence is thus considered derived within iguanines. Although *Conolophus* has enlarged, conical head scales overlying the nasal, prefrontal, and frontal bones similar to, yet smaller than, those seen in *Amblyrhynchus*, I have never observed osteoderms in *Conolophus*. The osteoderms of *Amblyrhynchus* are easily removed along with the skin, judging from their absence in most skeletal preparations of *Amblyrhynchus*, and it is therefore possible that *Conolophus* also possesses osteoderms. I will assume that osteoderms are absent in *Conolophus* until their presence is demonstrated.

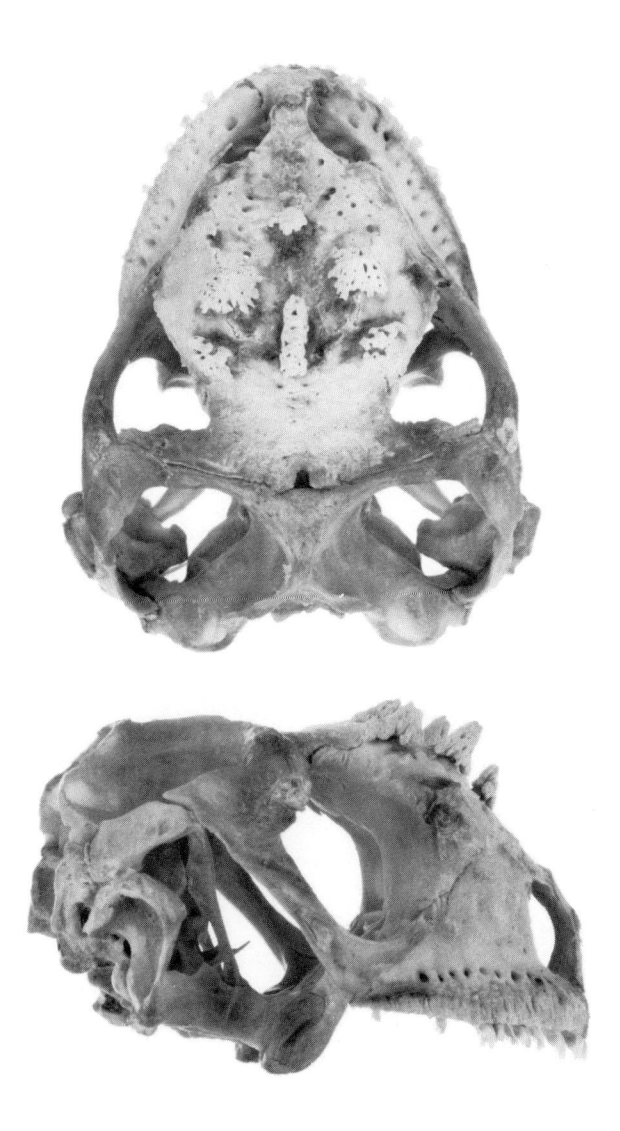

Plate 1. Dorsal (above) and lateral (below) views of the skull of *Amblyrhynchus cristatus* (JMS 127), showing osteoderms.

NONSKELETAL MORPHOLOGY

Iguanines exhibit considerable morphological variation in functional systems other than the skeleton, and I have therefore used certain nonskeletal characters for which relatively complete data on variation, both among all iguanine genera and for the four outgroups, were easily obtained. Characters in this section were taken from diagnoses in revisions, reviews, and faunal accounts as well as from the few comparative studies of nonskeletal anatomy of iguanines. I also include some obvious characters that I noticed in the course of this study.

ARTERIAL CIRCULATION

Zug (1971) was pessimistic about the systematic utility of the variation that he found in the patterns of the major arteries of iguanids. Nevertheless, I found at least three characters in his descriptions, as well as one additional character, that suggest monophyletic groups within Iguaninae. Other arterial characters may also be useful for phylogenetic studies within this taxon, but have not yet been studied in sufficient detail. Still other characters are either invariant among iguanines (e.g., branching pattern of the carotid arches, separation of the origins of dorsal aorta and subclavians) or variable within iguanine genera (e.g., separate origin of mesenterics versus origin from a common trunk), and thus cannot be used for examining relationships among these genera. These characters may be useful at different hierarchical levels.

It should be noted that Zug (1971) surveyed nearly all genera of Iguanidae, which limited him to relatively small samples for each genus (a maximum of four specimens for any iguanine genus). Zug did not examine *Conolophus*; my data are based on dissection of a single *C. subcristatus* (CAS 12058).

Zug reported that the subclavians of *Brachylophus* and *Dipsosaurus* are covered laterally by a thin, flat ligament, while those of other iguanines pass laterally beneath (=dorsal to?) a muscle bundle. My own observations on *Dipsosaurus* reveal muscle fibers in the thin sheets of tissue that cover the subclavians just lateral to their origins from the right systemic arch. Furthermore, whether muscular or ligamentous, the structures that cover the subclavians are the posterior portions of the paired *M. rectus capitis anterior* or their tendons, which originate on the ventral surfaces of the cervical vertebrae and insert on the exoccipitals and basioccipital lateral to the occipital condyle. Thus, even if the reported difference exists, it is a difference in the muscles rather than in the subclavian arteries.

The subclavians of *Conolophus* exhibit neither of the patterns described by Zug for other iguanines. In this taxon, the subclavians lie posterior and ventral to the origins of the *M. rectus capitis anterior* and are thus not covered by this muscle. For these reasons I use only the difference between the subclavians of *Conolophus* and those of all other iguanines as a systematic character.

According to Zug (1971), in *Dipsosaurus* and *Brachylophus* the dorsal aorta originates dorsal to the heart (by union of the left and right systemic arches), while in other iguanines it originates posterior to the heart. My observations on *Dipsosaurus* (n=1) and *Sauromalus* (n=1) reveal a profound difference supporting this distinction. In *Dipsosaurus* the systemic arches unite to form the dorsal aorta about as far posterior as the middle of the heart and the anterior end of the ninth vertebra. In *Sauromalus* the systemic arches remain paired much further posteriorly; they unite well behind the heart, near the middle of the 13th vertebra. *Conolophus*, however, is intermediate. The dorsal aorta in this taxon originates at about the level of the posterior end of the heart and the anterior end of the 10th vertebra. Because Zug did not discuss variation within his two categories, I arbitrarily placed *Conolophus* with those iguanines in which the dorsal aorta originates posterior to the heart.

Finally, I note minor exceptions to some of Zug's observations. In the single *Dipsosaurus* that I examined, the heart reaches the transverse axillary plane rather than being entirely anterior to this plane. In the single *Sauromalus* that I examined, the coeliac originates between, but separate from, the two mesenteric arteries.

COLIC ANATOMY

Iverson (1980) studied colic anatomy in iguanines. Variation within this group exists in the presence of colic valves, irregular colic folds, circular valves, semilunar valves, and in the number of colic valves. Although Iverson considered iguanine colic anatomy to be of limited phylogenetic value, at least two characters seem to be potentially useful for inferring phylogenetic relationships among iguanines. Nevertheless, because all of the colic modifications that characterize subsets of iguanines appear to be transformations of characters unique to iguanines, their polarity cannot be established by outgroup comparison until certain phylogenetic relationships within iguanines are determined. For example, one cannot use noniguanine outgroups to infer that colic folds are plesiomorphic relative to colic valves, or vice versa, because neither condition occurs in these outgroups.

The fact that noniguanines possess neither of the conditions found in iguanines is only a problem if these conditions are homologous members of a transformation series. Otherwise, each condition could be said to be lacking in the outgroups and therefore to be a separate apomorphic state. If they are homologous, however, one is forced to determine the apomorphy of the alternative conditions relative to each other. I assume homology between the colic valves and colic folds, because they share the common property of being infoldings of the same tissue components of the colic wall (Iverson, 1980). I also assume homology between circular and semilunar valves. The only difference between these two morphologies is whether or not the infolded tissue extends around the entire perimeter of the colon (Iverson, 1980). Because of the difficulties involved in outgroup comparison with the colic characters, I used them only at hierarchical levels less inclusive than Iguaninae as a whole.

Although much variation exists in the modal number of colic valves among iguanine taxa, this number is positively correlated with (maximum?) body size and does not change significantly during the postembryonic ontogeny of a given species (Iverson, 1980). Lack of a thorough study of the relationship between valve number and body size makes comparison of taxa that differ in body size problematic, and I have chosen not to use the numbers of different types of colic valves as systematic characters.

EXTERNAL MORPHOLOGY

Unlike the arterial and colic characters, which were obtained from comparative studies, the following characters were taken primarily from generic diagnoses or are based on personal observations. No adequate comparative descriptions of these characters exist in the literature, and I therefore describe them in more detail than the arterial and colic characters.

The scutellation of the iguanine head is complex and is potentially the source of many systematic characters. I note here only some obvious intertaxic differences and characters that have been used by previous authors.

Scales of the Snout and Dorsal Head. In most iguanines the snout terminates anteriorly in a median, azygous rostral scale. *Sauromalus* differs from all other iguanines in that it usually lacks an unpaired, median rostral (H. M. Smith, 1946: Fig. 38); the anteriormost snout scales above the lip are paired and separated by a median suture that meets the lip margin. According to Gates (1968), this character occurs in about 78% of *S. obesus*. All basiliscines, crotaphytines, morunasaurs, and oplurines possess a median, azygous rostral scale, indicating that the condition seen in *Sauromalus* is apomorphic within iguanines.

The other scales in the snout region also exhibit differences among iguanines. In most taxa they are relatively small, about the same size as the remaining dorsal cephalic scales. In *Iguana* and some *Cyclura*, however, these scales form large plates. Interspecific variation in this character is great within *Cyclura* (figures in Schwartz and Carey, 1977), ranging from the small scales much like those of other iguanines in *C. carinata*, *C. pinguis*, and *C. ricordii* to the large plates of *C. cychlura* and *C. nubila*. *Cyclura collei* and *C. rileyi* are intermediate, and the horns of *C. cornuta* are difficult to compare with the conditions seen in other taxa. Because outgroup comparison suggests that enlarged rostral scales are apomorphic (only *Laemanctus* among the outgroups examined has enlarged snout scales), either (1) the occurrence of this feature in *Iguana* and some *Cyclura* is convergent; (2) it indicates that *Iguana* is the sister group of some part of a paraphyletic *Cyclura*; or (3) enlarged snout scales is a synapomorphy of *Iguana* plus *Cyclura,* and some *Cyclura* have evolved small snout scales secondarily. Only a consideration of other characters can resolve this question.

Amblyrhynchus and *Conolophus* are similar to one another and differ from all other iguanines in the scalation of the dorsal surface of the head. In these two genera the dorsal head scales are pointed and conical, giving the head a rugose texture. This condition is more strongly developed in *Amblyrhynchus* than in *Conolophus*. All other iguanines have flat or only slightly domed dorsal head scales. In *Sauromalus hispidus* these scales are more strongly pointed than in the other taxa, but the condition is not nearly as extreme as in the Galápagos iguanas.

Like most iguanines, crotaphytines, oplurines, and most basiliscines have relatively flat head scales. *Laemanctus serratus* is the only basiliscine with conical head scales, but these scales are confined to the casque on the back of the head and do not extend onto the frontal and nasal regions as in the Galápagos iguanas. The dorsal head scales of morunasaurs are variable. In *Hoplocercus* and *Morunasaurus* these scales are convex but not pointed; in *Enyalioides* they are pointed and conical, but are relatively much smaller than those of the Galápagos iguanas. Thus, the condition of the dorsal head scales in *Amblyrhynchus* and *Conolophus* is not seen in any of the outgroups and must be considered apomorphic.

Superciliaries. Etheridge and de Queiroz (1988) noted variation in the superciliary scales of iguanines. In *Dipsosaurus* these scales are elongate anteroposteriorly and overlap one another extensively, especially in the anterior portion of the row. *Amblyrhynchus* and *Sauromalus* possess the opposite extreme in which the superciliaries are roughly quadrangular and nonoverlapping. The remaining iguanines are intermediate, with only moderate overlap of the superciliaries. Outgroup comparison indicates that the condition of the superciliaries has been relatively plastic at this level of comparison, making determination of its polarity ambiguous. Quadrangular, nonoverlapping superciliaries occur in morunasaurs and the basiliscine *Corytophanes*. Elongate, strongly overlapping superciliaries occur in oplurines, and an intermediate condition occurs in crotaphytines and the basiliscines *Basiliscus* and *Laemanctus*.

Suboculars. The morphology of the subocular scales is also variable in iguanines (Etheridge and de Queiroz, 1988). *Dipsosaurus* and *Ctenosaura* have one long and several shorter suboculars. In all other iguanines except *Amblyrhynchus*, which is intermediate, all of the suboculars are approximately equal in size. The condition of the suboculars in the four outgroups is too variable to allow inference about the polarity of this character. Basiliscines, morunasaurs, and some *Crotaphytus* have suboculars that are subequal in size. Other *Crotaphytus* have one moderately elongate subocular. *Gambelia* and oplurines have one very long subocular and several much shorter ones.

Anterior Auricular Scales (Van Denburgh, 1922). *Sauromalus* differs from all other iguanines in the scales that border the tympanum anteriorly, the anterior auricular scales. From two to five of these scales are enlarged relative to the neighboring scales and project posterolaterally over the tympanum, offering protection to this delicate membrane. In all other iguanines except *Dipsosaurus,* the anterior auricular scales are small and the tympanum is completely exposed. *Dipsosaurus* possesses a row of slightly enlarged anterior auricular scales. Outgroup comparison indicates that the enlarged anterior auriculars of *Sauromalus* are apomorphic. Basiliscines, *Crotaphytus*, *Hoplocercus*, *Morunasaurus*, and some *Enyalioides* lack enlarged anterior auricular scales, while in *Gambelia* and oplurines they are only slightly enlarged, roughly comparable to those of

Dipsosaurus. Some *Enyalioides* possess one or two seemingly nonhomologous large, pointed scales dorsal to the tympanum. Some sceloporines have anterior auriculars fully as large in proportion to their body size as those of *Sauromalus*; I consider this to be convergent.

Gular Region. All iguanines possess a transverse gular fold, although it is relatively weakly developed in *Amblyrhynchus* compared to other iguanines. A midsagittal gular expansion, or dewlap, is variably developed, but in no iguanine is it as highly extensible as in *Anolis*. A large dewlap is present in male *Brachylophus fasciatus* (Boulenger, 1885; Gibbons, 1981) and in both sexes of *B. vitiensis* (Gibbons, 1981), *Ctenosaura palearis* (Bailey, 1928), and *Iguana*. It is absent in *Amblyrhynchus*, *Conolophus*, most *Ctenosaura*, *Dipsosaurus*, and *Sauromalus*, but is weakly developed in *Cyclura* (Boulenger, 1885) and *Ctenosaura bakeri* (Bailey, 1928). The presence of a dewlap is not a simple dichotomy, as evidenced by the intermediate condition in *Cyclura* and *Ctenosaura bakeri*; nevertheless, a morphological gap exists between those taxa possessing a large dewlap and those in which it is weakly developed or absent.

A prominent gular fold occurs in all outgroup taxa used in this study and is, therefore, inferred to be plesiomorphic for iguanines. Although the absence of a dewlap is the most common condition among the outgroups, sufficient variation exists that this condition cannot be inferred to be plesiomorphic for iguanines as long as higher-level relationships remain unresolved. The dewlap is absent in *Basiliscus*, *Laemanctus*, crotaphytines, *Hoplocercus*, *Morunasaurus*, and oplurines, but it is present in *Corytophanes* and male *Enyalioides* (Boulenger, 1885).

Although a dewlap is developed to varying degrees in different iguanines, only the two species of *Iguana* possess a gular crest, a midsagittal row of enlarged scales extending below the throat along the edge of the dewlap. Because a gular crest is lacking in all outgroup taxa examined except *Corytophanes*, its presence in *Iguana* is inferred to be apomorphic.

Middorsal Scale Row. A row of scales aligned along the dorsal midline is present in all iguanines except *Sauromalus*. When present, the scales of the middorsal row are differentiated from the neighboring scales, although the degree of differentiation is highly variable. This variation ranges from the small, rounded knobs that form the row in *Dipsosaurus* to the tall curved spikes of large *Amblyrhynchus* and *Iguana*. In some *Cyclura* (Schwartz and Carey, 1977) and *Ctenosaura* (Bailey, 1928), the crest formed by the series of modified middorsal scales is interrupted in the shoulder or the sacral region. The presence of a middorsal scale row in the outgroups is highly variable, making it impossible to determine polarity at this level of analysis. A middorsal scale row is present in most basiliscines, *Enyalioides*, *Morunasaurus annularis*, and *Chalarodon*; it is absent in crotaphytines, *Laemanctus serratus*, *Morunasaurus groi*, *Hoplocercus*, and *Oplurus*.

Subdigital Scales of the Pes (Fig. 45). The conspicuous combs on the toes of *Cyclura* have long been used to diagnose this genus and especially to separate it from *Ctenosaura* (Barbour and Noble, 1916; Bailey, 1928; Schwartz and Carey, 1977). Similar toe denticulations, however, are known to occur in other iguanines (Gibbons, 1981). These

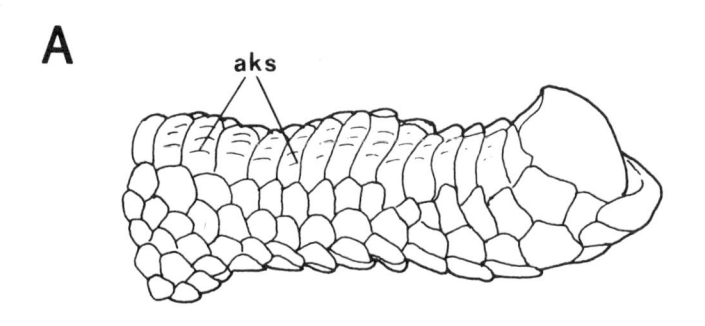

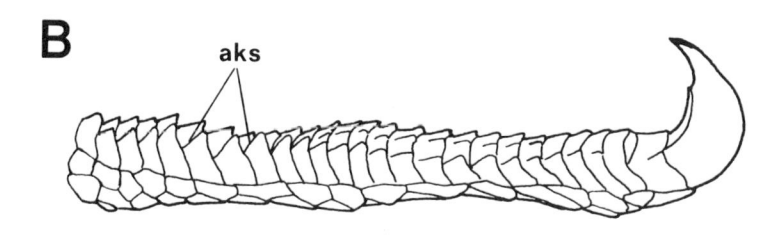

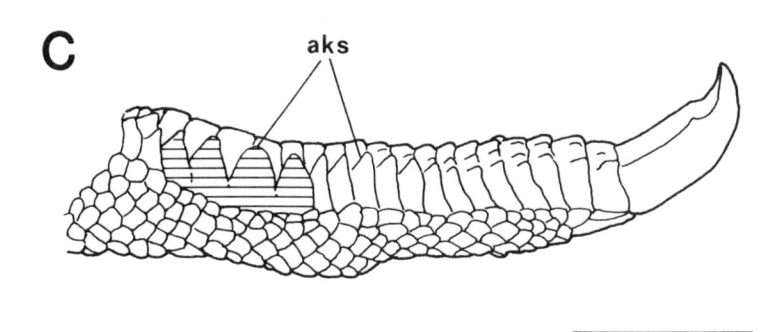

FIG. 45. Pedal digit II of (A) *Sauromalus obesus* (MVZ 35978), (B) *Brachylophus fasciatus* (CAS 54664), and (C) *Cyclura carinata* (CAS 54647) in anterodorsal view, showing differences in the morphology of the subdigital scales. Scale equals 1 cm. Fused subdigital scales are shaded. Abbreviations: aks, anterior keels of subdigital scales.

denticulations are formed by enlarged keels on the anterior edges of the subdigital scales. Varying degrees of enlargement of these keels are seen in iguanines. In *Sauromalus* the anterior keels of the subdigital scales are nearly the same size as the posterior ones (the subdigital scales are usually bi- or tricarinate), and the subdigital scales are roughly bilaterally symmetrical with respect to the long axis of the toe (Fig. 45A). In *Dipsosaurus* and *Iguana* the anterior keels of the subdigital scales are slightly larger than their posterior counterparts, and the subdigital scales are asymmetrical. Further enlargement of the anterior keels and a concomitant increase in the asymmetry of the pedal subdigital scales is seen in *Amblyrhynchus, Conolophus, Brachylophus* (Fig. 45B), and *Cyclura* (Fig. 45C) (increasing in size roughly in that order). Much of this variation can be seen within *Ctenosaura*.

All subdigital scales do not exhibit equal enlargement of the keels, which are usually largest under the first phalanx of digit II and the first and second phalanges of digit III. *Cyclura* and *Ctenosaura defensor* differ from other iguanines in that the scales bearing these largest keels are fused at their bases, giving the scales the appearance of a comb when viewed anteriorly (Fig. 45C). In *Cyclura* these combs are formed under the first phalanx of digit II and the first and second phalanges of digit III (illustrated in Barbour and Noble, 1916: Plates 13-15); in *Ctenosaura defensor* they occur only under the first phalanx of digit III.

Enlargement of the anterior keels of the subdigital scales is present in all outgroups examined in this study except basiliscines, though the degree of enlargement is variable. Basiliscines cannot be compared with iguanines because they have but a single median keel on the subdigital scales. In oplurines and crotaphytines the keels are moderately enlarged as in *Dipsosaurus*, but in morunasaurs (especially *Morunasaurus*) they are very large. Thus it is not possible to determine the precise plesiomorphic size of the keels of iguanines. Nevertheless, two conditions seen in iguanines can be considered to be apomorphic. Because the subdigital scales of all outgroups (except basiliscines) bear large anterior keels, the small anterior keels and concomitant symmetry of the subdigital scales in *Sauromalus* are apomorphic. Fusion of the bases of the subdigital scales with enlarged anterior keels is not seen in any outgroup and must also be considered apomorphic.

Hands and Feet. The hands and feet of *Amblyrhynchus* are partially webbed (Boulenger, 1885), which is presumably related to the semi-aquatic habits of these lizards and is unique among iguanids.

Caudal Squamation. One of the supposedly diagnostic features of *Ctenosaura* is a tail armed with strong, spinous scales (Bailey, 1928); however, similar caudal squamation also occurs in most *Cyclura* (Barbour and Noble, 1916; Schwartz and Carey, 1977). Within these two taxa the caudal squamation is highly variable among species. In some *Cyclura* (e.g., *C. cornuta*), the caudal scales in adjacent verticils are of similar size and are not spinous, a condition like that seen in most other iguanines. In the remaining *Cyclura* and in *Ctenosaura* the tail bears whorls of enlarged, spinous scales at regular intervals along its length. These whorls are separated by verticils of smaller scales that are smooth or much less spinous (except the middorsal scale row). The number of verticils between the whorls of enlarged, spinous scales is variable along the tail, generally decreasing posteriorly. The maximum number of rows between whorls of enlarged scales ranges from none in some *Ctenosaura defensor* (Bailey, 1928; Duellman, 1965) to about six in *Cyclura nubila* (Schwartz and Carey, 1977). Within *Ctenosaura*, there appears to be a negative correlation between the size of the scales in the enlarged whorls and both the number of scale rows between them and the relative length of the tail.

The evolution (or loss) of spinose tails appears to have occurred repeatedly within iguanids. Like most iguanines, basiliscines, crotaphytines, *Chalarodon*, and some *Enyalioides* have more or less uniform caudal squamation without spinous scales. Other *Enyalioides*, *Morunasaurus*, *Hoplocercus*, and *Oplurus* have whorls of enlarged spinous scales separated by smaller scales. The short, spinose tail of *Hoplocercus* is as extreme as anything seen in *Ctenosaura*. Although it seems likely that tails with whorls of enlarged, spinous scales are apomorphic within iguanines, this polarity is equivocal unless assumptions are made about either the relationships among outgroups and ingroup or those within morunasaurs and oplurines.

Cross-sectional Body Shape. *Sauromalus* differs from all other iguanines in its cross sectional body shape. All other iguanines are either laterally compressed or cylindrical in cross section, while *Sauromalus* is strongly depressed. The shape of the body of *Sauromalus* and several other of its distinctive skeletal features (e.g., low neural spines, horizontal orientation of the suprascapulae, short and broad pelvic girdle) are probably redundant characters. They are treated separately here because (1) the correlation among them is only hypothesized, and (2) some of them are known to change without accompanying changes in the others (e.g., not all depressed lizards have suprascapulae that form sharp angles with the scapulae).

Cross-sectional body shape in members of the four outgroups examined in this study varies in such a way that it is impossible to determine the plesiomorphic shape for iguanines. Basiliscines are laterally compressed. Some morunasaurs are compressed (*Enyalioides*) while others are depressed (*Hoplocercus*), and both crotaphytines and oplurines are depressed, though generally not as strongly as *Sauromalus*.

SYSTEMATIC CHARACTERS

Based on the descriptions of the iguanine skeleton and other anatomical features given above, I recognize the following systematic characters for use in phylogenetic analysis.

SKELETAL CHARACTERS

1. Ventral surface of premaxilla (Fig. 7): (A) bears large posterolateral processes; (B) posterolateral processes absent.
2. Posteroventral crests of premaxilla (Fig. 7): (A) small, do not continue up the sides of incisive process and are not pierced by foramina for maxillary arteries; (B) large, continue up sides of incisive process and are pierced or notched by foramina for maxillary arteries.
3. Anterior surface of rostral body of premaxilla: (A) broadly convex; (B) nearly flat.
4. Nasal process of premaxilla I (Figs. 6, 14, 45): (A) slopes backwards; (B) nearly vertical.
5. Nasal process of premaxilla II (Fig. 8): (A) wholly or partly exposed dorsally between nasals; (B) covered dorsally between nasals.
6. Size of nasals and nasal capsule (Figs. 5, 9, 11): (A) nasal capsule of moderate size, nasals relatively small; (B) nasal capsule enlarged, nasals relatively large.
7. Bones in anterior orbital region (Fig. 10): (A) lacrimal contacts palatine behind lacrimal foramen; (B) prefrontal contacts jugal behind lacrimal foramen.
8. Frontal (Figs. 5, 9, 11): (A) longer than wide, or length approximately equal to width; (B) wider than long.
9. Large paired openings at or near frontonasal suture: (A) absent; (B) present.
10. Cristae cranii on ventral surface of frontal (Fig. 12): (A) extend in a smooth continuous curve from frontal onto prefrontals; (B) frontal portions project anteriorly, forming a step between frontal and prefrontal portions.
11. Paired cristae on ventral surface of frontal medial to cristae cranii (Fig. 12): (A) absent or weakly developed; (B) strongly developed, united as a single median crest anteriorly and together with the cristae cranii forming pockets in the anteroventral surface of the frontal.
12. Dorsal borders of orbits (Figs. 5, 9, 11): (A) more or less smoothly curved; (B) wedge-shaped.
13. Position of parietal foramen (Figs. 5, 9, 11; Table 2): (A) on the frontoparietal suture; (B) variable (either A or C); or (C) within the frontal bone.

14. Supratemporals: (A) extend anteriorly more than halfway across the posterior temporal fossae; (B) extend anteriorly no more than halfway across the posterior temporal fossae.

15. Maxilla I: (A) relatively flat or concave laterally; (B) flares outward ventral to the row of supralabial foramina.

16. Maxilla II (Figs. 5, 14): (A) premaxillary process of maxilla lies roughly in the same plane as the remainder of the maxilla; (B) premaxillary process of maxilla curves dorsally.

17. Lacrimal: (A) large; (B) intermediate; (C) small.

18. Ventral process of squamosal (Fig. 15): (A) large; (B) small or absent.

19. Squamosal (Fig. 15): (A) separated from or barely contacting dorsal end of tympanic crest of quadrate; (B) abuts against dorsal end of tympanic crest of quadrate.

20. Septomaxilla: (A) flat, or with a weak ridge on anterolateral surface; (B) with a pronounced longitudinal crest.

21. Anterior dorsal surface of palatines (Fig. 16): (A) with a low medial ridge; (B) with a high medial crest.

22. Infraorbital foramen I (Fig. 17), process of palatine projecting posterolaterally or laterally behind the infraorbital foramen: (A) large; (B) small or absent.

23. Infraorbital foramen II (Fig. 17), process of palatine projecting posterolaterally or laterally behind the infraorbital foramen: (A) fails to contact jugal; (B) contacts jugal.

24. Infraorbital foramen III (Fig. 17): (A) located on the lateral or posterolateral edge of the palatine; (B) located entirely within the palatine (may or may not be connected by a suture to the lateral edge of the palatine).

25. Pterygoids (Figs. 5, 18): (A) medial borders relatively straight anterior to the pterygoid notch, pyriform recess narrows gradually; (B) medial borders curve sharply toward the midline anterior to the pterygoid notch, pyriform recess narrows abruptly.

26. Ectopterygoids: (A) fail to contact palatines near posteromedial corners of suborbital fenestrae; (B) usually contact palatines near posteromedial corners of suborbital fenestrae.

27. Parasphenoid rostrum (Fig. 20): (A) long; (B) short.

28. Cristae ventrolaterales of parabasisphenoid (Fig. 21): (A) strongly constricted behind basipterygoid processes; (B) intermediate; (C) widely separated.

29. Posterolateral processes of parabasisphenoid (Fig. 21): (A) present and large; (B) small or absent.

30. Laterally directed points on cristae interfenestralis: (A) absent; (B) present.

31. Stapes: (A) thin; (B) thick.

32. Relative heights of dorsal borders of dentary and surangular on either side of coronoid eminence (Fig. 22): (A) approximately equal; (B) dorsal border of dentary well above that of surangular.

33. Splenial: (A) large; (B) small.

34-35. Anterior inferior alveolar foramen (Fig. 23): (A) always between splenial and dentary, the coronoid may or may not contribute to its posterior margin; (B) entirely within the dentary in some specimens (others A); (C) between splenial and coronoid.

36. Labial process of coronoid (Fig. 24): (A) small; (B) intermediate; (C) large.

37. Angular I (Fig. 25): (A) extends far up the labial surface of the mandible and is largely visible in lateral view; (B) does not extend far up the labial surface of the mandible and is barely visible in lateral view.

38. Angular II: (A) wide posteriorly; (B) narrow posteriorly.

39. Surangular (Fig. 26): (A) exposed laterally only about as far forward as the apex of the coronoid or the anterior slope of this bone, and never anterior to the last dentary tooth; (B) exposed laterally well anterior to the apex of the coronoid and often anterior to the last dentary tooth.

40. Lingual exposure of surangular between ventral processes of coronoid (Fig. 27): (A) a dome-shaped portion exposed; (B) largely or completely covered by prearticular.

41. Angular process of prearticular (Fig. 28): (A) increases substantially in relative size during postembryonic ontogeny, becoming a prominent structure in adults; (B) increases only slightly in relative size during postembryonic ontogeny, remaining relatively small even in adults.

42. Retroarticular process (Figs. 28, 29): (A) tympanic and medial crests converge posteriorly to give the process a triangular outline in both juveniles and adults; (B) tympanic and medial crests converge posteriorly in juveniles, but the posterior ends separate during ontogeny so that the process assumes a quadrangular outline in adults.

43-44. Modal number of premaxillary teeth (Table 3): (A) fewer than seven; (B) seven; (C) more than seven.

45. Crowns of premaxillary teeth: (A) lateral cusps small or absent; (B) lateral cusps large.

46. Crowns of posterior marginal teeth I (Fig. 30): (A) tricuspid; (B) four-cusped; (C) polycuspate (5 to 10 cusps); (D) serrate.

47. Crowns of tricuspid posterior marginal teeth II (Fig. 30): (A) individual lateral cusps much smaller than apical cusp; (B) individual lateral cusps relatively large, subequal to apical cusp in size.

48. Pterygoid teeth I (Fig. 31): (A) entire row lies along the ventromedial edge of the pterygoid adjacent to the pyriform recess; B) posterior portion of row displaced laterally.

49. Pterygoid teeth II (Fig. 31): (A) entire row single throughout ontogeny; (B) posterior portion of row doubles ontogenetically; (C) entire row doubles ontogenetically.

50. Pterygoid teeth III (Fig. 31): (A) anterior portion of tooth patch present; (B) absent (posterior end of suborbital fenestra used as reference point).

51. Pterygoid teeth IV (Fig. 31): (A) usually present; (B) usually absent.

52-53. Hyoid I (Fig. 33): (A) second ceratobranchials short, often less than two-thirds the length of the first ceratobranchials; (B) intermediate, from two-thirds the length of the first ceratobranchials to slightly longer than the first ceratobranchials; (C) long, much longer than the first ceratobranchials.

54. Hyoid II (Fig. 33): (A) second ceratobranchials in medial contact with one another for most or all of their lengths; (B) separated from one another medially for most or all of their lengths.

55. Neural spines of presacral vertebrae (Figs. 34, 35): (A) tall, making up more than 50% of the total vertebral height; (B) short, making up less than 50% of the total vertebral height.

56. Zygosphenes (Fig. 36): (A) connected to prezygapophyses by a continuous arc of bone; (B) separated from zygapophyses by a deep notch.

57. Sacrum I: (A) posterolateral processes of second pleurapophyses (usually) present; (B) (usually) absent.

58. Sacrum II: (A) foramina in the ventral surfaces of the second pleurapophyses (usually) present; (B) (usually) absent.

59. Number of caudal vertebrae: (A) more than 40; (B) fewer than 40.

60. Autotomy septa in caudal vertebrae: (A) present (Fig. 37); (B) absent.

61. Beginning of the autotomic series of caudal vertebrae or beginning of the series of caudal vertebrae with two pairs of transverse processes (Fig. 37): (A) at or before the 10th caudal vertebra; (B) at or behind the 10th caudal vertebra.

62. Thin, midsagittal processes on the dorsal surface of the caudal centra anterior to the neural spines (Fig. 38): (A) relatively large and present well beyond the anterior third of the caudal sequence; (B) relatively small and confined to the anterior fifth of the caudal sequence.

63. Postxiphisternal inscriptional ribs: (A) do not form continuous chevrons (Fig. 39); (B) variably form continuous chevrons; (C) invariably form continuous chevrons.

64. Suprascapulae: (A) situated primarily in a vertical plane and forming a continuous arc with the scapulocoracoids; (B) situated primarily in a horizontal plane and forming an angle with the scapulocoracoids.

65. Scapular fenestrae (Fig. 40): (A) large, invariably present; (B) small or absent.

66. Posterior coracoid fenestrae (Fig. 40): (A) usually absent; (B) usually present.

67. Clavicles: (A) wide, with a prominent lateral shelf; (B) narrow, the lateral shelf small or absent.

68. Posterior process of the interclavicle (Fig. 40): (A) extends posteriorly beyond the lateral corners of the sternum; (B) does not extend beyond the lateral corners of the sternum.

69. Lateral processes of the interclavicle (Fig. 40): (A) usually forming angles of less than 75° with the posterior process and giving the interclavicle the shape of an arrow; (B) forming an angle of between 75° and 90° with the posterior process and giving the interclavicle the shape of a T.

70. Sternal fontanelle (Fig. 40): (A) present and of moderate size; (B) small or absent.

71. Sternum-xiphisternum (Fig. 40): (A) sternum diamond-shaped (quadrilateral), the xiphisterna in close proximity; (B) intermediate; (C) sternum pentagonal, the xiphisterna widely separated.

72. Pelvic girdle (Fig. 41): (A) long and narrow; (B) short and broad.

73. Anterior iliac process: (A) large; (B) small.
74. Osteoderms (Pl. 1): (A) absent; (B) present.

NONSKELETAL CHARACTERS

75. Heart (Zug, 1971): (A) does not extend posterior to the transverse axillary plane; (B) extends posterior to the transverse axillary plane.

76. Subclavian arteries (Zug, 1971; present study): (A) covered ventrally by the posterior end of the *M. rectus capitis anterior*; (B) not covered by the *M. rectus capitis anterior*.

77. Dorsal aorta (Zug, 1971): (A) right and left systemic arches unite to form the dorsal aorta above the heart; (B) origin of dorsal aorta posterior to heart.

78. Coeliac artery (Zug, 1971): (A) arises from the dorsal aorta anterior to and separate from the two mesenteric arteries; (B) arises posterior to the mesenterics, between the mesenterics, or continuous with one or the other of the mesenterics.

79. Colic wall (Iverson, 1980): (A) forms one or more transverse valves; (B) forms numerous irregular transverse folds.

80. Colic valves (Iverson, 1980): (A) all valves semilunar; (B) one or more valves circular (semilunar valves may be present or absent).

81. Rostral scale: (A) median and azygous; (B) subdivided by a median suture.

82. Scutellation of snout region: (A) consists of many small scales subequal in size to those of superorbital and temporal regions; (B) consists of relatively few large scales.

83. Dorsal head scales: (A) flat or slightly convex; (B) pointed and conical.

84. Superciliary scales (Etheridge and de Queiroz, 1988): (A) quadrangular and non-overlapping; (B) intermediate; (C) elongate and strongly overlapping.

85. Subocular scales (Etheridge and de Queiroz, 1988): (A) all subequal in size; (B) one or two suboculars moderately elongate; (C) one subocular very long, the rest shorter.

86. Anterior auricular scales: (A) all relatively small or one row slightly enlarged; (B) one row of scales anterior to tympanum pointed and greatly enlarged, extending posteriorly over tympanum.

87. Gular fold: (A) conspicuous; (B) weakly developed.
88. Dewlap: (A) small or absent; (B) present and large.
89. Gular crest: (A) absent; (B) present.
90. Middorsal scale row: (A) present; (B) absent.

91. Pedal subdigital scales I (Fig. 45): (A) anterior keels larger than posterior ones, scales asymmetrical; (B) anterior and posterior keels approximately equal in size, scales roughly symmetrical with respect to the long axis of the toe.

92. Pedal subdigital scales II (Fig. 45): (A) individual scales entirely separate; (B) scales with greatly enlarged anterior keels fused anteriorly at bases.

93. Toes: (A) unwebbed; (B) partially webbed.

94. Caudal squamation: (A) caudal scales in adjacent verticils approximately equal in size, smooth or keeled but not spinous; (B) tail bears whorls of enlarged, strongly spinous scales.

95. Cross-sectional body shape: (A) laterally compressed or cylindrical; (B) strongly depressed.

CHARACTER POLARITIES AND THE PHYLOGENETIC INFORMATION CONTENT OF CHARACTERS

Character-state distributions for the 95 characters among the four outgroups and the polarities inferred from these distributions are summarized in Table 5. Distributions of the characters among the basic taxa (genera) of iguanines are given in Table 6. Not surprisingly, the number of characters that exhibit variation within a basic taxon is correlated with the number of recognized species in the taxon.

Each character can be placed in one of four categories depending on its phylogenetic information content:

I. Unambiguous synapomorphies of basic taxa (characters 1, 2, 3, 4, 6, 9, 11, 12, 14, 15, 16, 17-2, 20, 22, 26, 27, 29, 30, 31, 32, 33, 34, 35, 36-2, 38, 41, 42, 46-3, 47, 49-2, 58, 64, 67, 72, 74, 75, 76, 81, 86, 87, 89, 91, 93). The derived condition of each of these characters is found in only one of the basic taxa and is characteristic of the taxon in which it is found. These characters support the monophyly of particular iguanine genera but provide no information about relationships among them.

II. Ambiguous synapomorphies of basic taxa (characters 10, 13-2, 24, 28-2, 53, 78, 82, 92). The derived condition of each of these characters is characteristic of one of the basic taxa but is also variably present in one or more other basic taxa. These characters are either (1) synapomorphies of one basic taxon that have arisen convergently in part of another one; (2) synapomorphies of one entire basic taxon plus part of another one that are indicative of the paraphyletic status of the latter; or (3) synapomorphies of a clade consisting of two or more basic taxa that have subsequently reversed within some of them. These characters may or may not provide information about relationships among basic taxa.

III. Derived characters shared by two or more basic taxa (characters 5, 7, 8, 13, 17, 18, 19, 21, 23, 25, 28, 36, 37, 39, 40, 45, 46, 46-2, 48, 50, 51, 52, 54, 62, 66, 68, 69, 70, 77, 83). The derived condition of each of these characters is characteristic of more than one of the basic taxa and may or may not occur variably in one or more of the others. These characters are the primary data relevant to an analysis of relationships among the basic taxa. Because of character incongruence, the interpretation of these characters as synapomorphies is not always straightforward, and a reasonable interpretation of any one character must take the others into consideration. Some of the similarities are undoubtedly homoplastic and must ultimately be interpreted as more than one synapomorphy.

IV. Characters of undeterminable polarity (characters 43, 44, 55, 56, 57, 59, 60, 61, 63, 65, 71, 73, 79, 80, 84, 85, 88, 90, 94, 95). These characters are too variable either within or among the outgroups, or both, for any reasonable inference to be made about their polarity. Therefore, these characters cannot be used as evidence for phylogenetic relationships within Iguaninae until either the relationships of the outgroups to iguanines are determined (Maddison et al., 1984) or some phylogenetic structure within iguanines is established so that some iguanines can serve as outgroups to others in an analysis of a less inclusive group (Watrous and Wheeler, 1981).

TABLE 5. Distributions of Character States of 95 Characters Among Four Outgroups to Iguanines and the Polarities That Can Be Inferred From Them

Character	Outgroup				Polarity Inference
	Ba	Cr	Mo	Op	
1	A	A	A	A	A=0, B=1
2	A	A	A	A	A=0, B=1
3	A	A	A	A	A=0, B=1
4	A	A	A	A	A=0, B=1
5	A	A[1]	A	A	A=0, B=1
6	A	A	A	A	A=0, B=1
7	A	B	A	A	A=0, B=1
8	A	A	A	A	A=0, B=1
9	A	A	A	A	A=0, B=1
10	A	A	A	A	A=0, B=1
11	A	A	A	A	A=0, B=1
12	A	A	A	A	A=0, B=1
13	A,C	A	A,N	A	A=0, B=1, C=2
14	A	A	A	A	A=0, B=1
15	A	A	A	A	A=0, B=1
16	A	A	A	A	A=0, B=1
17	A	A	C[2]	A	A=0, B=1, C=2
18	A,B	A	A	A	A=0, B=1
19	A	A	A	A	A=0, B=1
20	A	A	A	A	A=0, B=1
21	A	A	A	A	A=0, B=1
22	A	A	A	A,B	A=0, B=1
23	A	A,B[3]	A	A	A=0, B=1
24	A	A	A	A	A=0, B=1
25	A	B	A	A	A=0, B=1

TABLE 5 (continued)

Character	Outgroup Ba	Outgroup Cr	Outgroup Mo	Outgroup Op	Polarity Inference
26	A	A	A	A	A=0, B=1
27	A	A	A	A	A=0, B=1
28	A	A	A	A	A=0, B=1, C=2
29	A	A,B	A	A	A=0, B=1
30	A	A	A	A	A=0, B=1
31	A	A	A	A	A=0, B=1
32	A	A	A	A	A=0, B=1
33	A	A	A	A	A=0, B=1
34	A,X	A,X	A,X	A,X	A=0, B=1, C=0
35	A,X	A,X	A,X	A,X	A=0, B=0, C=1
36	A	A	A	A	A=0, B=1, C=2
37	A	A	A	A,B	A=0, B=1
38	A	A	A	A,B	A=0, B=1
39	A	A	A	A	A=0, B=1
40	A,X	A	A	A,B	A=0, B=1
41	A,B	A	A	A	A=0, B=1
42	A	A	A,B	A	A=0, B=1
43	B,C	A,B	A,B[4],C	A	?
44	B,C	A,B	A,B[4],C	A	?
45	A	A	A	A	A=0, B=1
46	A	A	A,C	A	A=0, B=1, C=2, D=3
47	A	A	A	A	A=0, B=1
48	A	A	A	A	A=0, B=1
49	A	A	A	A	A=0, B=1, C=2
50	A	A	A	A	A=0, B=1
51	A	A	A	A	A=0, B=1
52	B	A	B	B	A=1, B=0, C=0

TABLE 5 (continued)

Character	Outgroup				Polarity Inference
	Ba	Cr	Mo	Op	
53	B	A	B	B	A=0, B=0, C=1
54	A	A	A	A	A=0, B=1
55	A	B	A,B	A,B	?
56	B	A	A,B	A	?
57	B	A,B	A,B	A,B	?
58	A	A	A	A	A=0, B=1
59	A	A	A,B	A,B	?
60	A,B	A,B	A,B	A	?
61	A,B,N	B,N	A,N	A	?
62	B	A	A	A	A=0, B=1
63	A	A	C	C[5]	?
64	A	A	A	A	A=0, B=1
65	B	A	A,B	A,B[6]	?
66	A	B	A	A	A=0, B=1
67	A	A	A	A,B	A=0, B=1
68	A	A,B	A	A	A=0, B=1
69	A,B	A	A	A	A=0, B=1
70	A	A	A	A	A=0, B=1
71	A	A,B	A	A,C	?
72	A	A	A	A	A=0, B=1
73	A,B	A	B	A	?
74	A	A	A	A	A=0, B=1
75	A	A	A[7]	A[8]	A=0, B=1
76	A	A	-	-	A=0, B=1
77	A	A	A[7]	A,B	A=0, B=1
78	A,X	A	B[7]	A	A=0, B=1

TABLE 5 (continued)

Character	Outgroup Ba	Cr	Mo	Op	Polarity Inference
79	X	X	X	X	?
80	X	X	X	X	?
81	A	A	A	A	A=0, B=1
82	A,B	A	A	A	A=0, B=1
83	A,X	A	A,B	A	A=0, B=1
84	A,B	B	A	C	?
85	A	A,B,C	A	C	?
86	A	A	A	A	A=0, B=1
87	A	A	A	A	A=0, B=1
88	A,B	A	A,B	A	?
89	A,B	A	A	A	A=0, B=1
90	A	B	A,B	A,B	?
91	A,X	A	A	A	A=0, B=1
92	A	A	A	A	A=0, B=1
93	A	A	A	A	A=0, B=1
94	A	A	A,B	A,B	?
95	A	B	A,B	B	?

Note: Abbreviations: Ba, basiliscines; Cr, crotaphytines; Mo, morunasaurs; Op, oplurines. Character-state codes correspond with those given in the list of systematic characters, with the following additions: N, not applicable; X, character state not found in iguanines; -, no data. Polarity codes are as follows: 0, plesiomorphic; 1, apomorphic; 2, derived from 1; 3, derived from 2.
[1]Some intermediate between A and B.
[2]Lacrimal absent in some.
[3]Probably not homologous with condition A in iguanines (see text).
[4]Sample too small to determine mode.
[5]Descriptively condition A but thought to be derived from C for other reasons (see text).
[6]Descriptively condition B but may be derived from A (see text).
[7]*Hoplocercus* not examined.
[8]Intermediate.

TABLE 6. Distributions of Character States of 95 Characters Among Eight Iguanine Taxa

Taxon	Character																				
	1	2	3	4	5	6	7	8	9	10	11	12	13	14	15	16	17	18	19	20
Amblyrhynchus	1	0	1	1	1	1	1	1	0	0	1	1	0	0	1	0	2	0	0	1
Brachylophus	0	0	0	0	0	0	0	0	0	0	0	0	0	0	0	0	0	1	0	0
Conolophus	0	1	0	0	1	0	1	1	0	0	0	0	0	1	0	0	1	1	0	0
Ctenosaura	0	0	0	0	0	0	0,1	0	0	0,1	0	0	0,1	0	0	1	0	1	0	0
Cyclura	0	0	0	0	0	0	0,1	0,1	0	0	0	0	0,1,2	0	0	0	0	1	1	0
Dipsosaurus	0	0	0	0	0	0	0	0	1	0	0	0	2	0	0	0	0	1	0	0
Iguana	0	0	0	0	0	0	0	0,1	0	0	0	0	0	0	0	0	0	0	1	0
Sauromalus	0	0	0	0	0	0	0	0	0	0	0	0	1	0	0	0	0	1	0	0

TABLE 6 (continued)

Taxon	21	22	23	24	25	26	27	28	29	30	31	32	33	34	35	36	37	38	39	40
Amblyrhynchus	1	0	1	0,1	1	0	1	0	0	0	1	1	0	0	1	1	1	0	0	1
Brachylophus	0	0	1	1	0	0	0	0	0	0	0	0	0	1	0	1	0	0	0	0
Conolophus	1	0	1	0	1	1	0	0	0	0	0	0	0	0	0	2	1	0	0	1
Ctenosaura	0	0	1	0,1	1	0	0	0	1	0	0	0	0	0	0	0	0	0	0,1	0
Cyclura	0	0	1	0	1	0	0	1,2	0	0	0	0	0	0	0	0	0	0	1	0,1
Dipsosaurus	0	1	0	0	1	0	0	0	0	1	0	0	0	0	0	0	0	0	0	0
Iguana	0	0	1	0	1	0	0	2	0	0	0	0	0	0	0	0	0	0	1	0
Sauromalus	0	0	0,1	0,1	1	0	0	0	0	0	0	0	1	0	0	0	1	1	0	0

TABLE 6 (continued)

Taxon	41	42	43	44	45	46	47	48	49	50	51	52	53	54	55	56	57	58	59	60
Amblyrhynchus	1	0	B	B	1	0	1	1	0	1	0	1	0	1	A	B	A	0	A	B
Brachylophus	0	0	B	B	0	0,1	0	0	0	0	0	0	1	0	A	B	A	0	A	B
Conolophus	0	0	B	B	1	1	0	1,N	N	1	1	1	0	0	A	B	A	1	A	B
Ctenosaura	0	0	B	B	0	0,1,2	0	1	0,1	0	0	0	0	0	A	B	B	0	A,B	A
Cyclura	0	0	C	C	0	2	0	1	0,1	0	0	0	0	0	A	B	A,B	0	A	A
Dipsosaurus	0	1	B	B	0	1	0	0,N	0,N	1	1	0	0	0	A	A	A	0	A	A
Iguana	0	0	B	B	0	3	0	1	2	0	0	0	0,1	0	A	B	B	0	A	A,B
Sauromalus	0	0	A	A	0	2	0	1	0	0	0	1	0	1	B	B	A	0	B	A

Character

TABLE 6 (continued)

Taxon	61	62	63	64	65	66	67	68	69	70	71	72	73	74	75	76	77	78	79	80
Amblyrhynchus	B	0	B	0	B	1	0	1	1	1	B	0	A	1	0	0	1	0	B	N
Brachylophus	A	1	C	0	A	0	0	0	0	0	A	0	A	0	0	0	0	0	A	A
Conolophus	B	0	B	0	A	1	0	0	0	0	A	0	A	0	0	1	1	0	A	B
Ctenosaura	B	0	B	0	A	1	0	0	0	0	A	0	A	0	0	0	1	0	A	B
Cyclura	B	0	B	0	A	1	0	0	0	0	A	0	A	0	0	0	1	0	A	B
Dipsosaurus	A	0	A	0	A	0	0	0	0	0	A	0	A	0	0	0	0	0	A	B
Iguana	B	1	B	0	A	1	0	0	0	0	A	0	A	0	0	0	1	1	A	B
Sauromalus	B	0	A	1	B	1	1	1	1	1	C	1	B	0	1	0	1	0,1	A	B

TABLE 6 (continued)

Taxon	81	82	83	84	85	86	87	88	89	90	91	92	93	94	95
Amblyrhynchus	0	0	1	A	B	0	1	A	0	A	0	0	1	A	A
Brachylophus	0	0	0	B	A	0	0	B	0	A	0	0	0	A	A
Conolophus	0	0	1	B	A	0	0	A	0	A	0	0	0	A	A
Ctenosaura	0	0	0	B	C	0	0	A,B	0	A	0	0,1	0	B	A
Cyclura	0	0,1	0	B	A	0	0	A	0	A	0	1	0	A,B	A
Dipsosaurus	0	0	0	C	C	0	0	A	0	A	0	0	0	A	A
Iguana	0	1	0	B	A	0	0	B	1	A	0	0	0	A	A
Sauromalus	1	0	0	A	A	1	0	A	0	B	1	0	0	A	B

Note: Polarities are based on data in Table 5. Abbreviations: 0, plesiomorphic; 1, apomorphic; 2, transformation of 1; 3, transformation of 2; N, not applicable. Character states of characters whose polarities were undeterminable using basiliscines, crotaphytines, morunasaurs, and oplurines as outgroups are indicated by alphabetic codes corresponding with those in the character list.

ANALYSIS OF PHYLOGENETIC RELATIONSHIPS

Following character analysis, the phylogenetic relationships among the eight iguanine genera (basic taxa) and the diagnoses of various monophyletic groups of iguanines, including the basic taxa, were determined by means of a three-step procedure: (1) First, the derived characters shared by the members of two or more basic taxa were used for a preliminary analysis of phylogenetic relationships. (2) Second, certain phylogenetic relationships based on the preliminary analysis were used to identify new outgroups for the determination of polarities of characters that were undeterminable using basiliscines, crotaphytines, morunasaurs, and oplurines as outgroups. These characters were then added to the existing set for a second analysis of relationships within a subgroup of iguanines. (3) Third, the results of the two analyses were combined to produce a final estimate of phylogenetic relationships within Iguaninae, and the level at which each character exists as a synapomorphy was reanalyzed in light both of these relationships and of variation within the basic taxa.

PRELIMINARY ANALYSIS

A total of 29 characters (numbers 5, 7, 8, 13, 17, 18, 19, 21, 23, 25, 28, 36, 37, 39, 40, 45, 46, 48, 50, 51, 52, 54, 62, 66, 68, 69, 70, 77, 83), representing a minimum of 30 phylogenetic transformations, were used in the preliminary analysis of relationships. These are the characters whose derived states are shared by two or more basic taxa (category III). Character 46 has multiple states, with two levels (states 1, 2, and 3; states 2 and 3) that are shared by two or more taxa. This accounts for the difference between the number of characters and the minimum number of phylogenetic transformations. Table 7 gives the 29 characters rescored to eliminate variation within basic taxa, as described under Materials and Methods, above.

The results of the preliminary analysis are summarized in Figure 46. Two different cladograms (Fig. 46A,B) can account for the distribution of derived characters with a minimum number of phylogenetic character transformations. In terms of the phylogenetic relationships suggested, the two cladograms differ only in the positions of *Brachylophus* and *Dipsosaurus*. In one (Fig. 46A), *Dipsosaurus* is the sister group of all other iguanines; in the other (Fig. 46B), *Brachylophus* occupies this position instead.

The minimum-step cladograms require 46 phylogenetic character transformations (consistency index = 0.65), 16 more than the absolute minimum of 30, which would only obtain if all of the characters had mutually compatible distributions among the basic taxa.

TABLE 7. Distributions of Character States of 29 Characters Used in the Preliminary Analysis

Taxon	\multicolumn{15}{c}{Character}
	5	7	8	13	17	18	19	21	23	25	28	36	37	39	40
Amblyrhynchus	1	1	1	0	1	0	0	1	1	1	0	1	1	0	1
Brachylophus	0	0	0	0	0	1	0	0	1	0	0	1	0	0	0
Conolophus	1	1	1	0	1	1	0	1	1	1	0	1	1	0	1
Ctenosaura	0	0	0	0	0	1	0	0	1	1	0	0	0	0	0
Cyclura	0	0	0	0	0	1	1	0	1	1	1	0	0	1	0
Dipsosaurus	0	0	0	1	0	1	0	0	0	1	0	0	0	0	0
Iguana	0	0	0	0	0	0	1	0	1	1	1	0	0	1	0
Sauromalus	0	0	0	1	0	1	0	0	0	1	0	0	1	0	0
CI (Fig. 46A)	1.00	1.00	1.00	0.50	1.00	0.33	1.00	1.00	0.50	0.50	1.00	0.50	1.00	1.00	1.00
CI (Fig. 46B)	1.00	1.00	1.00	0.50	1.00	0.33	1.00	1.00	0.33	1.00	1.00	0.50	1.00	1.00	1.00

Taxon	\multicolumn{14}{c}{Character}
	45	46	48	50	51	52	54	62	66	68	69	70	77	83
Amblyrhynchus	1	0	1	1	0	1	1	0	1	1	1	1	1	1
Brachylophus	0	0	0	0	0	0	0	1	0	0	0	0	0	0
Conolophus	1	1	1	1	1	1	0	0	1	0	0	0	1	1
Ctenosaura	0	0	1	0	0	0	0	0	1	0	0	0	1	0
Cyclura	0	2	1	0	0	0	0	0	1	0	0	0	1	0
Dipsosaurus	0	1	0	1	1	0	0	0	0	0	0	0	0	0
Iguana	0	2	1	0	0	0	0	1	1	0	0	0	1	0
Sauromalus	0	2	1	0	0	1	1	0	1	1	1	1	1	0
CI (Fig. 46A)	1.00	0.40	1.00	0.50	0.50	1.00	0.50	0.50	1.00	0.50	0.50	0.50	1.00	1.00
CI (Fig. 46B)	1.00	0.40	1.00	0.50	0.50	1.00	0.50	0.50	1.00	0.50	0.50	0.50	1.00	1.00

Note: Characters have been recoded so as to eliminate variation within basic taxa as well as derived states that characterize single taxa. Consistency indices (CI) for the characters on the two minimum-step cladograms based on these characters (Figs. 46A and B) are also given.

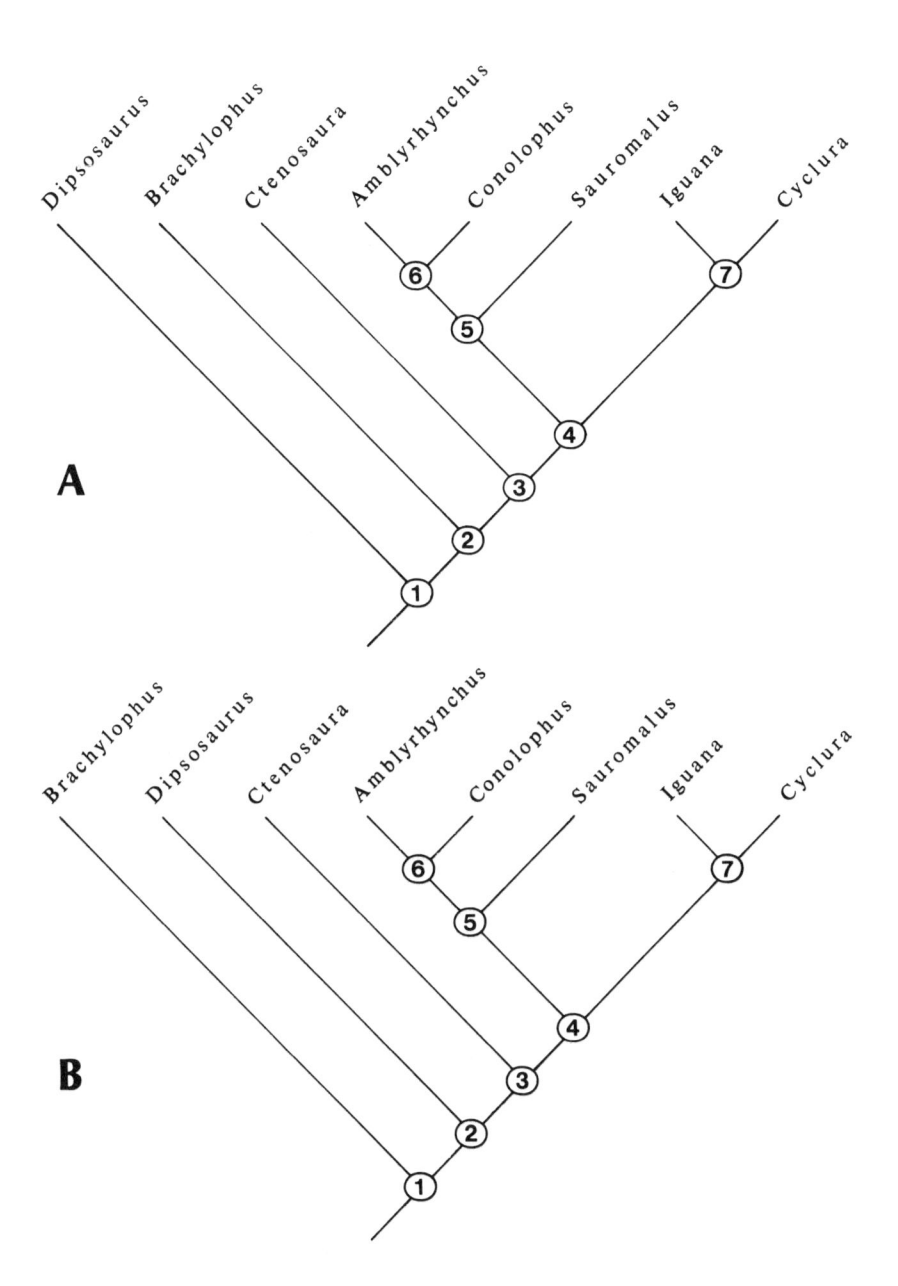

FIG. 46. Minimum-step cladograms for eight basic taxa of iguanines, resulting from a preliminary analysis of 29 characters (Table 7). Two different cladograms (A and B) account for the taxic distribution of derived characters with 46 character transformations. Synapomorphies of the numbered nodes and basic taxa are given in the text.

The consistency indices (Kluge and Farris, 1969) for each of the characters on each of the two minimum-step cladograms are given in Table 7. The C-index is a measure of the deviation of a character from a perfect fit (C-index of 1.00) to a given cladogram. Synapomorphies for the various nodes of the cladograms are given below by the number of the character and the letter of the character state as designated in the list of systematic characters. Convergent characters are underlined; characters involving reversal are marked with an asterisk. Because only characters whose derived states are shared by two or more of the basic taxa were used in this analysis, any character interpreted as a synapomorphy of a basic taxon necessarily exhibits homoplasy.

Figure 46A: Node 1: 18-B*, 25-B*; Node 2: 23-B*; Node 3: 48-B, 66-B, 77-B; Node 4: 46-B*, 46-C or-D*; Node 5: 37-B, 52-53-A; Node 6: 5-B, 7-B, 8-B, 17-B or-C, 21-B, 36-B or-C, 40-B, 45-B, 46-B or-A*, 50-B, 83-B; Node 7: 19-B, 28-B or-C, 39-B; *Amblyrhynchus*: 18-A*, 46-A*, 54-B, 68-B, 69-B, 70-B; *Brachylophus*: 25-A*, 36-B, 62-B; *Conolophus*: 51-B; *Ctenosaura*: none; *Cyclura*: none; *Dipsosaurus*: 13-C, 46-B, 50-B, 51-B (last two characters are redundant); *Iguana*: 18-A*, 62-B; *Sauromalus*: 13-B, 23-A*, 54-B, 68-B, 69-B, 70-B.

The synapomorphies of the second cladogram (Fig. 46B) are identical to those of the first (Fig. 46A), with the following exceptions: Node 1: 18-B*, 23-B*; Node 2: 25-B, 46-B,-C, or-D*; Node 4: 46-C or-D*; *Brachylophus*: 36-B, 62-B; *Ctenosaura*: 46-A*.

Six of the homoplastic characters on the first minimum-step cladogram (Fig. 46A) can be interpreted in more than one way, each involving the same number of phylogenetic transformations. These alternative interpretations are diagrammed in Figure 47. Character 25-B can be interpreted as convergent synapomorphies of *Dipsosaurus* on the one hand and of all other iguanines except *Brachylophus* (node 3) on the other hand (Fig. 47A). Alternatively, it can be interpreted as a synapomorphy of all iguanines that has reversed in *Brachylophus* (Fig. 47B). Characters 54-B, 68-B, 69-B, and 70-B can be interpreted as convergent synapomorphies of *Amblyrhynchus* on the one hand and of *Sauromalus* on the other (Fig. 47C). Alternatively, these characters can be interpreted as synapomorphies of the Galápagos iguanas plus *Sauromalus* (node 5) that have reversed in *Conolophus* (Fig. 47D). Two alternative interpretations of character 46 are diagrammed in Figure 47E and F. Both interpretations require five phylogenetic transformations.

Alternative interpretations of homoplastic characters on the second minimum-step cladogram (Fig. 46B) are identical to those on the first (Fig. 46A), with the following exceptions: Character 25-B has only one possible minimum-step interpretation; it is a synapomorphy of all iguanines except *Brachylophus* (node 2). Character 23-B can either be interpreted as convergent synapomorphies of *Brachylophus* on the one hand and the taxa united above node 3 (Fig. 48A) on the other hand, or it can be interpreted as a synapomorphy of all iguanines that has subsequently reversed in *Dipsosaurus* (Fig. 48B). The same alternative interpretations of character 46 are available for the second minimum-step cladogram as for the first, but two additional alternatives exist (Fig. 48C,D).

Of the six subterminal nodes on each of the two minimum-step cladograms resulting from the preliminary analysis, three (nodes 3, 6, and 7) are well supported. That is, these

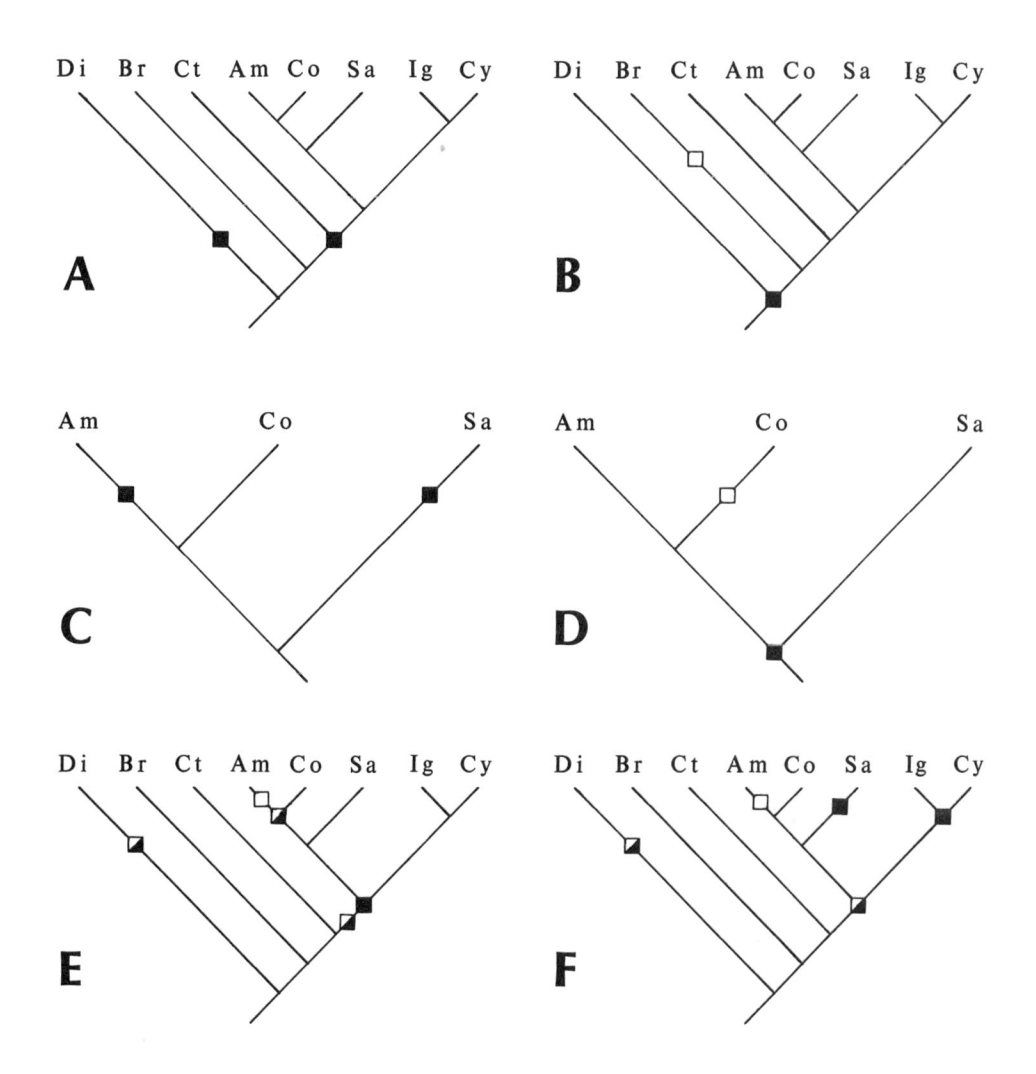

FIG. 47. Alternative interpretations of character transformation for homoplastic characters on a minimum-step cladogram (Fig. 47A). A and B are alternative interpretations for character 25; C and D for characters 54, 68, 69, and 70; E and F for character 46. Solid squares represent transformations to the derived condition; open squares represent reversals; half-solid squares represent intermediate states.

nodes are diagnosed by more than two derived characters that are unique and unreversed and strongly outweigh conflicting characters. Node 1 is also well supported, but it is supported by the results of an analysis at a more inclusive hierarchical level. Node 2 is the most weakly supported, for it supports the monophyly of different groups of basic taxa on the two minimum-step cladograms.

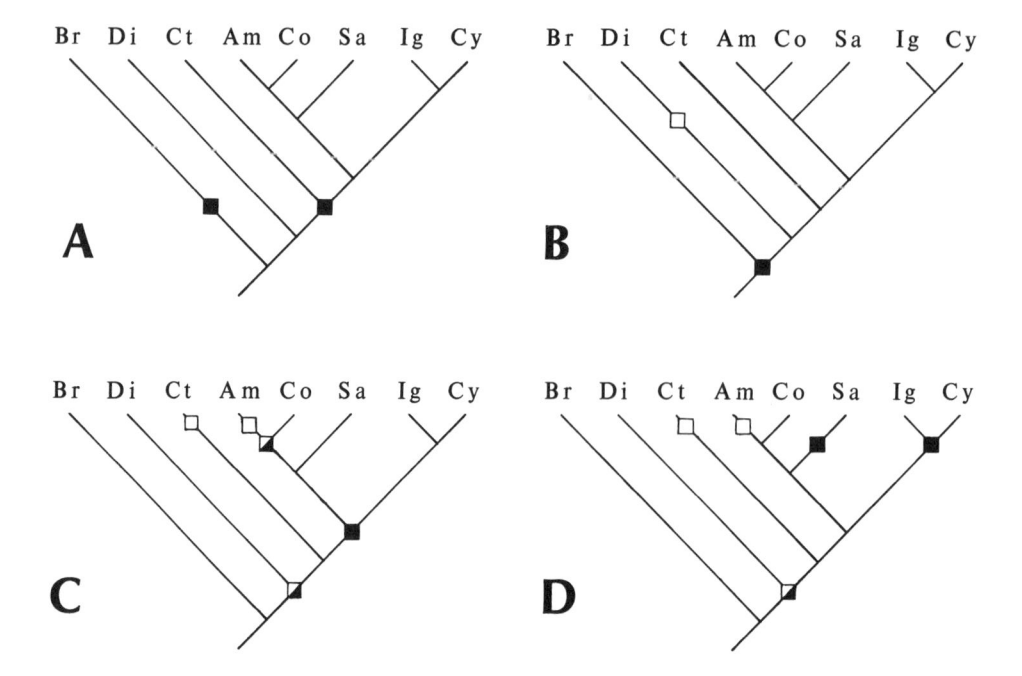

FIG. 48. Alternative interpretations of character transformation for homoplastic characters on a minimum-step cladogram (Fig. 47B). A and B are alternative interpretations for character 23; C and D for character 46. Solid squares represent transformations to the derived condition; open squares represent reversals; half-solid squares represent intermediate states.

LOWER-LEVEL ANALYSIS

In an attempt to gain better resolution of iguanine phylogenetic relationships, I performed an analysis at a lower hierarchical level (node 3), using *Brachylophus* and *Dipsosaurus* as outgroups in order to determine the polarities of characters that were undeterminable at the level of all iguanines. I chose node 3 for this analysis because it is the most inclusive group within iguanines whose monophyly is well supported.

The precise relationships of *Brachylophus* and *Dipsosaurus* to the rest of the iguanines are problematical. One of the minimum-step cladograms resulting from the preliminary analysis has *Dipsosaurus* as the sister group of all other iguanines (Fig. 46A), while the other has *Brachylophus* in this position instead (Fig. 46B). The second hypothesis might at first appear to be better supported, because *Dipsosaurus* shares two derived characters with the other iguanines (characters 25-B and 46-B,-C, or-D), while *Brachylophus* shares only one derived character (23-B) with them. However, character 46 has four equally simple alternative interpretations, and in only two of these (Fig. 48C,D) does it support a sister-group relationship between *Dipsosaurus* and all iguanines other than *Brachylophus*.

Under the other two alternative interpretations, the presence of the first derived state in *Dipsosaurus* is considered to be convergent, as in Figure 47E and F. For this reason, I have chosen to leave the relationships among *Brachylophus*, *Dipsosaurus*, and the new ingroup (node 3) unresolved in the assessment of polarities for the lower-level analysis.

I used the same basic methodology for determining polarities in the lower-level analysis (Appendix III) that I used in the preliminary analysis, where the relationships of the outgroups to the ingroup are uncertain (Appendix II). For reasons presented in Appendix III, I considered polarity to be determinable only when both *Brachylophus* and *Dipsosaurus* exhibit the same character state.

Using *Brachylophus* and *Dipsosaurus* as additional outgroups for analysis at a lower hierarchical level, I was able to determine polarities for 13 of the 20 characters whose polarities could not initially be determined (Table 8). The number of premaxillary teeth turns out to be two characters (hence the numbering in the character list as characters 43-44) representing transformations in opposite directions from the ancestral condition, a mode of seven premaxillary teeth. Although character 84 (superciliary scales) differs in *Brachylophus* and *Dipsosaurus*, it seems reasonable to conclude that state A is derived, since it is found in neither *Brachylophus* nor *Dipsosaurus* and represents one end of a continuum that has the conditions seen in these two taxa at the other end. Both *Brachylophus* and *Dipsosaurus* exhibit the same state for character 61, but this character is irrelevant to an analysis of relationships at the level in question because it does not vary within the new ingroup. Characters 56 and 80 also do not vary within the ingroup, but their polarities are undeterminable because *Brachylophus* and *Dipsosaurus* exhibit different conditions.

The use of *Brachylophus* and *Dipsosaurus* as additional outgroups for an analysis of relationships at a lower hierarchical level necessitates a reevaluation of the polarities of those characters whose polarities had already been determined using more remote outgroups. The reasoning behind polarity reevaluation is similar to that behind polarity assessment and is presented in Appendix IV. Under this reasoning, the only characters whose polarity assessments needed to be changed after reevaluation were character 18 (polarity reversed) and character 46 (changed to undeterminable). Character 46 is a four-state character, and what becomes undeterminable is whether state A or state B is ancestral. Therefore, I have lumped states A and B as state 0 and consider states C and D to be successively more derived conditions (i.e., C = 1, D = 2).

Eight of the characters used in the preliminary analysis of relationships among all iguanines cannot be used in the analysis of relationships of all iguanines other than *Brachylophus* and *Dipsosaurus*, either because they must be interpreted as synapomorphies of a basic taxon that are convergent with a condition found in *Brachylophus* or *Dipsosaurus* (characters 13, 51, and 62) or because they do not vary within the new ingroup (characters 25, 48, 66, and 77). These characters were removed from consideration, and the remaining characters were combined with those whose polarities were newly determined, using *Brachylophus* and *Dipsosaurus* as outgroups, and whose derived states characterized

TABLE 8. Polarity Inferences for Lower-level Analysis Using *Brachylophus* and *Dipsosaurus* as Outgroups

Character	Outgroup *Brachylophus*	Outgroup *Dipsosaurus*	Polarity Inference
43	B	B	A=1, B=0, C=0
44	B	B	A=0, B=0, C=1
55	A	A	A=0, B=1
56	B	A	?[1]
57	A	A	A=0, B=1
59	A	A	A=0, B=1
60	B	A	?
61	A	A	?[1]
63	C	A	?
65	A	A	A=0, B=1
71	A	A	A=0, B=1, C=2
73	A	A	A=0, B=1
79	A	A	A=0, B=1
80	A	B	?[1]
84	B	C	A=1; B,C=0
85	A	C	?
88	B	A	?
90	A	A	A=0, B=1
94	A	A	A=0, B=1
95	A	A	A=0, B=1

Note: Character-state codes correspond with those in the list of systematic characters. Polarity codes are as follows: 0, plesiomorphic; 1, apomorphic; 2, derived from 1.
[1]Character does not vary in the ingroup.

more than one of the basic taxa. The data set for this lower-level analysis consists of 26 characters, and is presented with the characters recoded to eliminate variation within the basic taxa (Table 9).

TABLE 9. Distributions of Character States of 26 Characters Among Six Taxa Within a Subset of Iguaninae

Taxon	Character												
	5	7	8	17	18	19	21	23	28	36	37	39	40
Amblyrhynchus	1	1	1	1	1	0	1	1	0	1	1	0	1
Conolophus	1	1	1	1	0	0	1	1	0	1	1	0	1
Ctenosaura	0	0	0	0	0	0	0	1	0	0	0	0	0
Cyclura	0	0	0	0	0	1	0	1	1	0	0	1	0
Iguana	0	0	0	0	1	1	0	1	1	0	0	1	0
Sauromalus	0	0	0	0	0	0	0	0	0	0	1	0	0
CI	1.00	1.00	1.00	1.00	0.50	1.00	1.00	0.50	1.00	1.00	1.00	1.00	1.00

Taxon	Character												
	45	46	50	52	54	57	65	68	69	70	71	83	84
Amblyrhynchus	1	0	1	1	1	0	1	1	1	1	1	1	1
Conolophus	1	0	1	1	0	0	0	0	0	0	0	1	0
Ctenosaura	0	0	0	0	0	1	0	0	0	0	0	0	0
Cyclura	0	1	0	0	0	0	0	0	0	0	0	0	0
Iguana	0	1	0	0	0	1	0	0	0	0	0	0	0
Sauromalus	0	1	0	1	1	0	1	1	1	1	1	0	1
CI	1.00	0.50	1.00	1.00	0.50	0.50	0.50	0.50	0.50	0.50	0.50	1.00	0.50

Note: Characters have been recoded to eliminate variation within basic taxa as well as derived states that characterize single basic taxa (not including *Brachylophus* and *Dipsosaurus*). Consistency indices (CI) for the characters are also given; these are identical for the three minimum-step cladograms based on the characters in this table (Fig. 49A,B,C) and their consensus cladogram (Fig. 50).

Three fully resolved cladograms of equal and minimum length can be constructed from the 26 characters used in the lower-level analysis (Fig. 49). These cladograms differ only in the position of *Ctenosaura*, which in turn depends on the interpretation of character 57, the presence or absence of posterolaterally directed processes on the pleurapophyses of the second sacral vertebra. The derived absence of these processes occurs in *Ctenosaura*, *Iguana*, and some *Cyclura*, but was scored absent for the latter taxon in order to simplify analysis. This is one of only two derived characters out of the set of 26 that occurs invariably in *Ctenosaura* and is relevant to the placement of this taxon within the restricted ingroup. The only other derived character that occurs invariably in *Ctenosaura* (character 23-B) also occurs in all ingroup taxa except some *Sauromalus*. Therefore, provided that *Sauromalus* is monophyletic, this character is most reasonably interpreted as a synapomorphy of the entire ingroup that has reversed in some *Sauromalus*. If the sister-group relationship between *Iguana* and *Cyclura*, based on other characters, is accepted, then character 57-B might be interpreted as convergent in *Iguana* on the one hand and in *Ctenosaura* on the other. If so, *Ctenosaura* can have any of the relationships illustrated in Figure 49; given this information alone, there is no reason to prefer any one of these alternative placements over the others. Alternatively, character 57-B might be interpreted as a synapomorphy of a clade consisting of *Ctenosaura*, *Iguana*, and *Cyclura* that has subsequently reversed within *Cyclura*. Because *Cyclura* is actually variable for this character, the hypothesis of acquisition and reversal requires fewer phylogenetic transformations than does that of convergence (two instances versus three). Although one of the three cladograms (Fig. 49A) would be favored under such an interpretation, the difference is so small that little importance can be attached to it in terms of resolving the placement of *Ctenosaura*. Therefore, I consider the relationships of *Ctenosaura* within the restricted ingroup to be uncertain.

Because the three minimum-step cladograms resulting from the lower-level analysis differ only in the placement of *Ctenosaura*, I present diagnostic synapomorphies for a single consensus cladogram (Adams, 1972) that leaves the relationships of *Ctenosaura* unresolved (Fig. 50). This consensus cladogram is identical to the other three in terms of evolutionary steps, requiring 37 phylogenetic character transformations out of the absolute minimum of 26 (C-index = 0.70), which would only obtain if all characters had compatible distributions among basic taxa. The consistency indices (Kluge and Farris, 1969) for the characters on the consensus cladogram (Fig. 50) are identical to those on the three minimum-step cladograms (Fig. 49A,B,C) from which it was derived. These are given in Table 9. Synapomorphies for the nodes of the consensus cladogram (Fig. 50) are given below, with convergent characters underlined and characters involving reversal marked with an asterisk.

Node 1: 23-B*; Node 2: 37-B, 52-A; Node 3: 5-B, 7-B, 8-B, 17-B or -C, 21-B, 36-B or-C, 40-B, 45-B, 50-B, 83-B; Node 4: 19-B, 28-B or-C, 39-B, <u>46-C or-D</u>; *Amblyrhynchus*: <u>18-A, 54-B, 65-B, 68-B, 69-B, 70-B, 71-B, 84-A</u>; *Conolophus*: none; *Ctenosaura*: <u>57-B</u>; *Cyclura*: none; *Iguana*: <u>18-A, 57-B</u>; *Sauromalus*: 23-A*, <u>46-C, 54-B, 65-B, 68-B, 69-B, 70-B, 71-C, 84-A</u>.

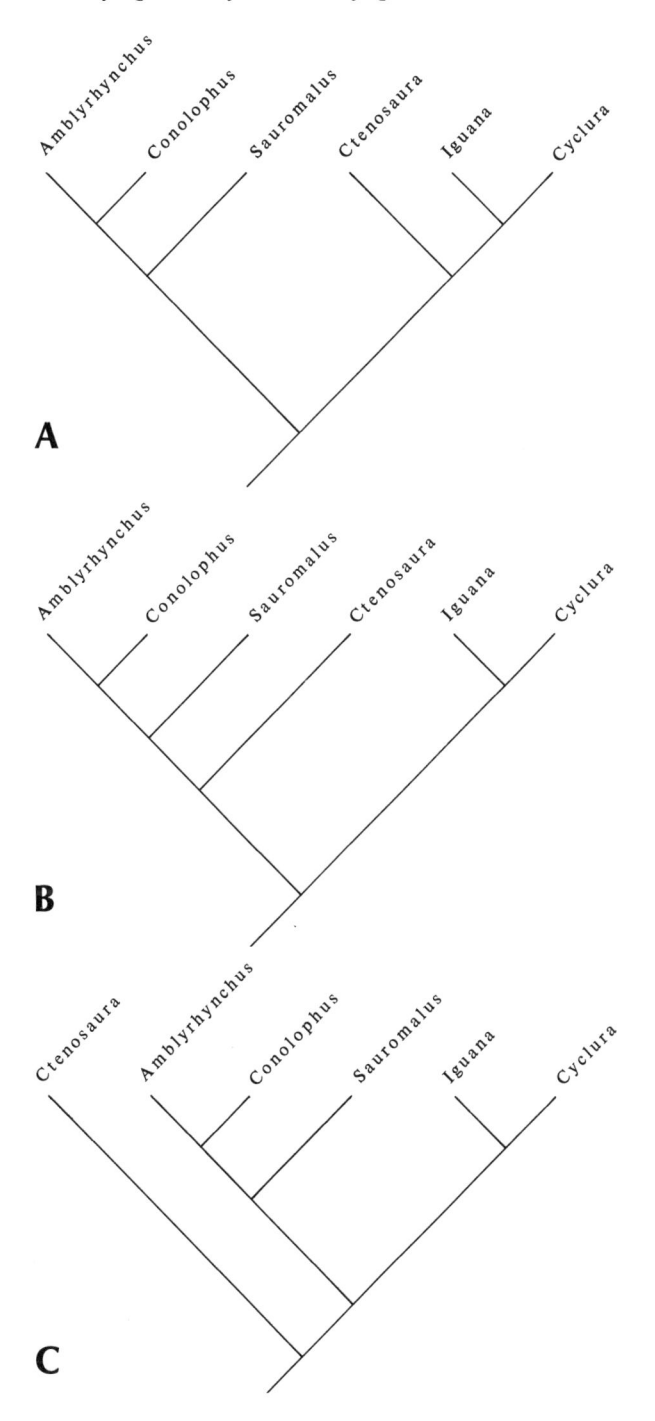

FIG. 49. Minimum-step cladograms resulting from an analysis of 26 characters (Table 9) in a subset of iguanines. Three different cladograms (A, B, and C) account for the taxic distribution of derived characters with 37 character transformations.

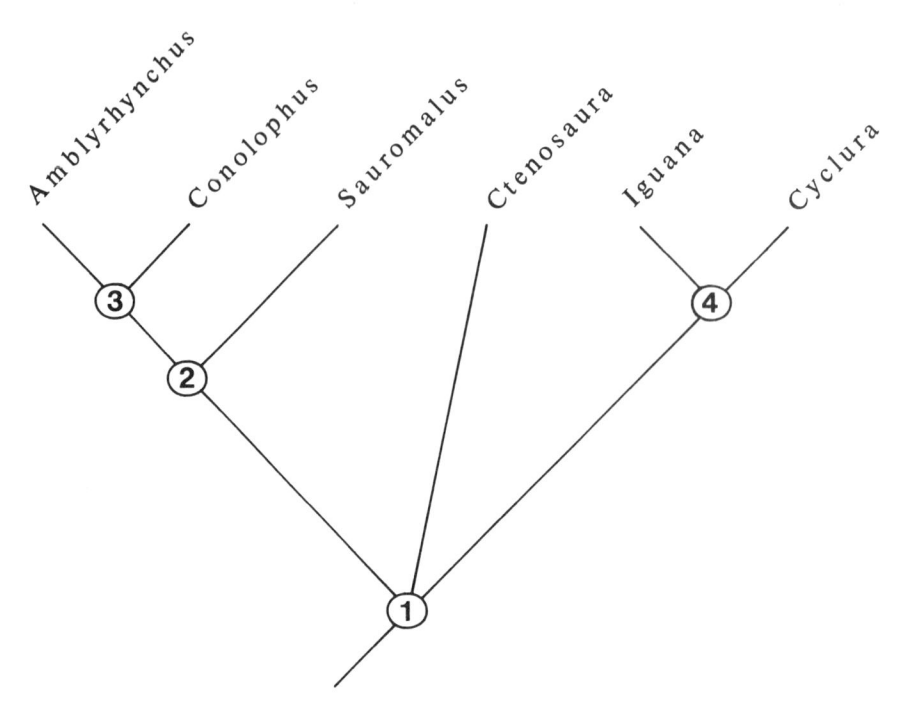

FIG. 50. Consensus cladogram for the three cladograms illustrated in Figure 49. The consensus cladogram is also a minimum-step cladogram in that it requires the same number of character transformations as do the three fully resolved cladograms upon which it is based. Synapomorphies for the numbered nodes and the basic taxa are given in the text.

Eight of the eleven homoplastic characters can be interpreted in two different ways, each involving the same number of phylogenetic transformations on the minimum-step cladograms. The alternative interpretations of character 57 have already been discussed. Its derived state is either convergent in *Ctenosaura* and *Iguana*, or it is a synapomorphy of a monophyletic group composed of *Ctenosaura*, *Iguana*, and *Cyclura* that has subsequently reversed in *Cyclura*. Characters 54, 65, 68, 69, 70, 71, and 84 are either convergent in *Amblyrhynchus* and *Sauromalus* or they are synapomorphies of a monophyletic group composed of *Amblyrhynchus*, *Conolophus*, and *Sauromalus* that have subsequently reversed in *Conolophus*.

Although all three of the subterminal nodes on the consensus cladogram (not including node 1, which is a conclusion of a higher-level analysis) are supported by at least two derived characters, every one is contradicted by some other characters. Node 2, suggesting a sister-group relationship between *Sauromalus* and the Galápagos iguanas, is supported by two characters: reduced labial exposure of the angular bone (37-B) and short second ceratobranchials (52-53-A). Nevertheless, the possession of polycuspate or serrate marginal tooth crowns (character 46-B or-C) suggests that *Sauromalus* is more closely related to *Iguana* and *Cyclura*, while the lack of lateral contact between palatine and jugal posterior to the infraorbital foramen (character 23-A) suggests that *Sauromalus* may be the sister group of all other iguanines in the lower-level analysis. However, this character is actually variable within *Sauromalus* and may have reversed within this taxon.

Node 4, suggesting a sister-group relationship between *Iguana* and *Cyclura*, is supported by four characters: squamosal abuts against dorsal end of quadrate (19-B); cristae ventrolateralis of parabasisphenoid relatively widely separated (28-B or-C); surangular extends far forward on lateral surface of mandible (39-B); and polycuspate or serrate marginal tooth crowns (46-C or-D). One of these characters (46) actually suggests monophyly of a more inclusive group consisting of *Sauromalus*, *Iguana*, and *Cyclura*. Another character, absence of posterolateral processes on pleurapophyses of second sacral vertebra (character 57-B), suggests a sister-group relationship between *Iguana* and *Ctenosaura*, although most *Cyclura* also lack the processes. Yet another character, large ventral process of the squamosal (18-A), suggests a sister-group relationship between *Amblyrhynchus* and *Iguana* (the homology of this character is dubious but cannot be ruled out on morphological grounds alone).

Node 3, suggesting a sister-group relationship between *Amblyrhynchus* and *Conolophus*, is the best-supported node. It is diagnosed by 10 derived characters: nasal process of premaxilla covered dorsally between nasals (5-B); prefrontal contacts jugal behind lacrimal foramen (7-B); frontal wider than long (8-B); reduction of lacrimal (17-B or-C); medial crest on anterior dorsal surface of palatine (21-B); enlarged labial foot of coronoid (36-B or-C); surangular covered lingually below coronoid (40-B); premaxillary teeth with large lateral cusps (45-B); anterior portion of pterygoid tooth patch absent (50-B); and pointed, conical dorsal head scales (83-B). Nevertheless, seven derived characters suggest a sister-group relationship between *Amblyrhynchus* and *Sauromalus*: medial separation of second ceratobranchials (54-B); reduction or loss of scapular fenestrae (65-B); short posterior process of interclavicle (68-B); T-shaped interclavicle (69-B); reduction or loss of sternal fontanelle (70-B); medial separation of xiphisterna (71-B or-C); and quadrangular, nonoverlapping superciliary scales (84-A). *Conolophus* lacks all of these derived characters. Therefore, if a sister-group relationship between *Amblyrhynchus* and *Conolophus* is accepted, then the derived characters shared by *Amblyrhynchus* and *Sauromalus* must either be convergent or reversed in *Conolophus*.

PHYLOGENETIC CONCLUSIONS

PREFERRED HYPOTHESIS OF RELATIONSHIPS

Figure 51 summarizes my conclusions about phylogenetic relationships among the genera of iguanine lizards, based on the two analyses discussed above as well as a consideration of variation within basic taxa. Synapomorphies of the various taxa are given in the Diagnoses section, below. Although this is not the most fully resolved cladogram that can be obtained from the characters used in this study, it indicates the best-supported monophyletic groups. The differences between this cladogram and the most fully resolved cladogram that can be obtained from these data are as follows: (1) Either *Brachylophus* or *Dipsosaurus* can be considered the sister group of all other iguanines on a fully resolved cladogram. Since both hypotheses are equally reasonable in terms of the characters discussed here, I leave the relationships among *Brachylophus*, *Dipsosaurus*, and the monophyletic group composed of all other iguanines unresolved. (2) Although it is possible to place *Ctenosaura* as the sister group of the clade composed of *Iguana* and *Cyclura*, this conclusion is based on one of two possible interpretations of a single character, and this character must later be lost within the clade that it is supposed to diagnose. I prefer to leave the relationships of *Ctenosaura* to *Sauromalus*, *Iguana* and *Cyclura*, and *Amblyrhynchus* and *Conolophus* unresolved. (3) Finally, a fully resolved cladogram places *Sauromalus* as the sister group of the Galápagos iguanas, while I leave the relationships of *Sauromalus* to *Ctenosaura*, the Galápagos iguanas, and *Iguana* and *Cyclura* unresolved. The reasons for these differences are discussed more fully in the sections on phylogenetic analysis, above, and the diagnoses of the monophyletic groups of iguanines, below.

CHARACTER EVOLUTION WITHIN IGUANINAE

Although the primary goal of this study was to determine the relationships among the genera of iguanine lizards, I was only partially successful in this endeavor. Other than Iguaninae as a whole, I recognize only three monophyletic groups composed of more than one of the basic taxa, whereas a fully resolved dichotomously branching phylogeny would have six such groups. Failure to resolve relationships cannot be attributed to a lack of morphological variation within Iguaninae, for derived characters-which are sometimes numerous-support the monophyly of each of the basic taxa. Therefore, it seems that most of the character evolution within iguanines occurred after the lineages leading to the extant

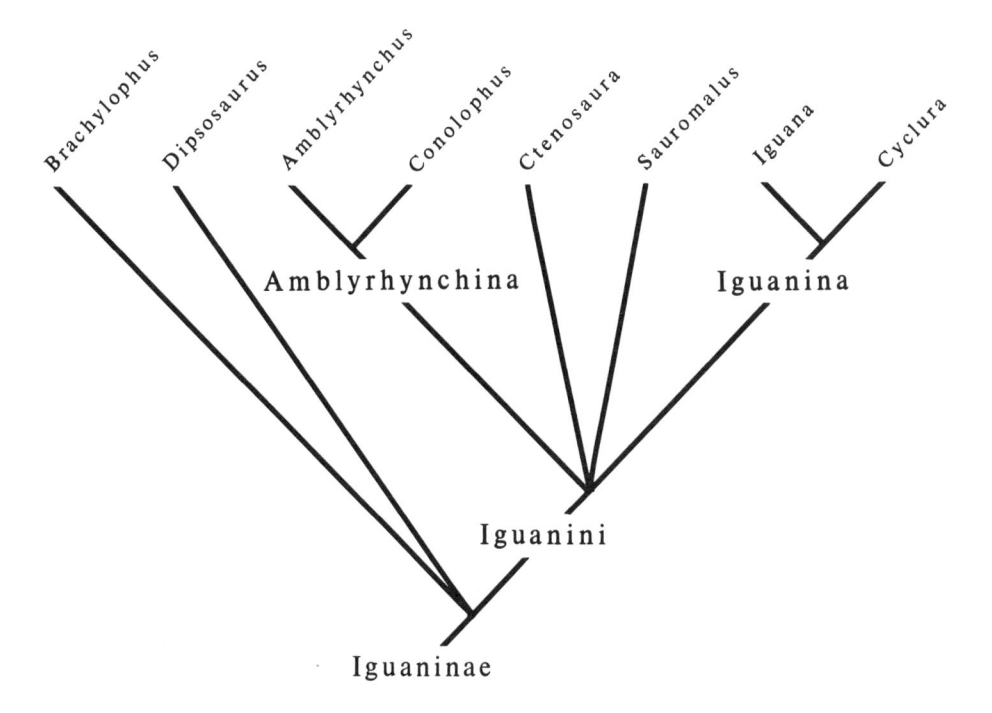

FIG. 51. Phylogenetic relationships within Iguaninae according to the present study.

genera had already diverged from one another. Accepting this proposition might lead one to conclude that these lineages separated during a relatively brief time interval and that they have been evolving separately for a long time. Implicit in this conclusion, however, is the assumption that rates of character evolution are similar in separately evolving lineages. This assumption is contradicted by the distribution of derived characters among the basic taxa and the relationships that can be resolved by them. For example, *Amblyrhynchus* possesses more obvious derived characters not found in *Conolophus* than does either *Brachylophus* or *Dipsosaurus*, even though *Conolophus* apparently shared a more recent common ancestor with *Amblyrhynchus* than it did with either *Brachylophus* or *Dipsosaurus*. Given that the characters used in this study are representative of overall phenotypic evolution, one must conclude that the lineage leading to *Amblyrhynchus* has evolved more rapidly than those leading to *Brachylophus* and *Dipsosaurus*.

COMPARISONS WITH PREVIOUS HYPOTHESES

Although a close relationship among some or all of the taxa currently placed in Iguaninae was recognized by several nineteenth-century authors, no explicit hypotheses about phylogenetic relationships among the various iguanine genera appeared until the twentieth century. The phylogenetic relationships proposed here are both similar in some respects and different in others when compared with previous hypotheses about iguanine relationships. In this section, I evaluate these previous hypotheses in light of the results of the present study.

Barbour and Noble (1916) and Bailey (1928) both hypothesized a close relationship between *Cyclura* and *Ctenosaura*, and Schwartz and Carey (1977) further proposed that *Cyclura* originated from *Ctenosaura*. Neither of these hypotheses is supported by the results of the present study. First, *Ctenosaura* possesses at least three characters that are derived relative to the condition seen in *Cyclura* (premaxillary process of maxilla curves dorsally; short posterolateral processes of parabasisphenoid; elongate subocular scale), and thus cannot be considered ancestral to the latter. Second, *Cyclura* shares more derived characters with *Iguana* than it does with *Ctenosaura*, implying that *Cyclura* shared a more recent common ancestor with *Iguana* than with *Ctenosaura*. The relationships among these three taxa are discussed further in the comments on *Cyclura* in the Diagnoses section, below.

Mittleman (1942) proposed a phylogenetic scheme for the North American iguanids, including *Ctenosaura*, *Dipsosaurus*, and *Sauromalus* (Fig. 1). This phylogeny was modified slightly by H. M. Smith (1946), who removed *Ctenosaura* from a position of direct ancestry to all other North American iguanids and placed *Dipsosaurus* and *Sauromalus* close to a group composed of what are now considered the sceloporines and crotaphytines rather than to just part of this radiation (compare Figs. 1 and 2). Although Smith did not include iguanines other than those occurring within or very near to the United States in his branching diagram, it is clear from his comments on the "herbivore section" (group II in Fig. 2) that he also considered *Iguana*, *Amblyrhynchus*, *Conolophus*, and *Cyclura* to be part of this group.

Common to the Mittleman (1942) and Smith (1946) phylogenies is the notion that iguanines are ancestral to the other North American iguanids-that is, that some iguanines shared a more recent common ancestor with these other iguanids than they did with other iguanines. This idea seems to be related to another notion held by both Mittleman and Smith, namely that iguanines are "primitive" iguanids. According to Mittleman (1942:112), "*Dipsosaurus* is probably the most primitive of the North American *Iguanidae* (excepting *Ctenosaura*, which is properly a Central and South American form)." H. M. Smith (1946:101) says of his herbivore section (iguanines), "this includes the large, primitive iguanids."

The notions that iguanines are "primitive" iguanids and that they are ancestral to sceloporines and crotaphytines are false. While it is true that iguanines lack certain derived features seen in these other groups, this is simply a manifestation of the mosaic nature of evolution, for the converse is also true. Sceloporines and crotaphytines lack derived characters seen in iguanines. Iguanines are derived relative to sceloporines and crotaphytines in numerous characters, among them the possession of caudal vertebrae with two pairs of transverse processes, the posterior location of the supratemporal bone, herbivory and associated morphological adaptations (flared tooth crowns, colic valves), and large body size. Because some of the derived characters of iguanines occur nowhere else within Iguanidae, iguanines cannot be considered ancestral to any other iguanids.

In the early 1960's, Etheridge constructed a phylogeny for iguanines as part of his scheme of relationships for the entire Iguanidae (Fig. 4). This scheme was never intended to be published (Etheridge, pers. comm.), and it is difficult to evaluate because the reasons for the various groupings were not specified. Other than differences in resolution, the results of the present study differ from Etheridge's scheme in two primary ways: While I consider *Dipsosaurus* and *Brachylophus* to be outside of a monophyletic group formed by the remaining iguanines, Etheridge considered *Brachylophus* to be the sister group of the Galápagos iguanas, and he considered *Dipsosaurus* to be the sister group of *Sauromalus*. Although the relationships proposed by Etheridge can be supported by particular shared, derived characters (e.g., lack of autotomy septa in caudal vertebrae of *Brachylophus* and the Galápagos iguanas; anterior position of parietal foramen in *Sauromalus* and *Dipsosaurus*), the weight of the evidence suggests different relationships and necessitates that the distribution of these derived characters is partly the result of convergence. The full evidence leading to this conclusion is given in the diagnoses of the various monophyletic groups recognized in the present study and will not be repeated here.

The only published study dealing with relationships among all the iguanine genera is that of Avery and Tanner (1971). As I noted in the Introduction, these authors used an artificial system for assessing similarity, used many characters that are probably correlated, made no attempt to determine character polarity, and did not specify how their similarity data were used to construct their phylogenetic tree. Furthermore, Avery and Tanner's conclusions are obscured by self-contradictory, vague, and ambiguous statements. For example, they state (p. 69) that "the osteological characters ... indicate that *Oplurus* and *Chalarodon* are more closely related to each other than to the iguanines, and *Oplurus* is the Madagascarian genus most closely related to the Western Hemisphere iguanines." In one place (p. 68), Avery and Tanner claim that *Ctenosaura* is certainly ancestral to the Western Hemisphere iguanines, but their phylogenetic tree (Fig. 3) suggests that *Dipsosaurus*, a Western Hemisphere iguanine, is not derived from *Ctenosaura*, and later (p. 73) they seem to consider *Ctenosaura* ancestral to only *Cyclura* and *Sauromalus*. One of Avery and Tanner's 11 numbered conclusions is that *Iguana* and *Ctenosaura* evolved from a common ancestral stock. This statement is uninformative, for they consider all iguanines to have evolved from a common ancestor; it is also misleading when compared with their phylogenetic tree (Fig. 3). For these reasons, I find it impossible to compare my conclusions with those of Avery and Tanner.

Wyles and Sarich (1983) published the results of immunological comparisons for 10 species of iguanines representing all eight genera. Given the limitations of these data, their results are in general agreement with the relationships proposed here. Wyles and Sarich's comparisons are incomplete in that antisera were prepared to only four of the iguanine species, and immunological distances to all other iguanines in the study are given for the antisera to only two of the four, *Amblyrhynchus* and *Conolophus*. Assuming that immunological distance is roughly proportional to time of divergence, Wyles and Sarich's data suggest (1) that *Amblyrhynchus* and *Conolophus* are sister taxa; (2) that the Galápagos iguanas are roughly equally closely related to *Ctenosaura*, *Cyclura*, *Iguana*, and *Sauromalus*; and (3) that they are more distantly related to *Dipsosaurus* and *Brachylophus*. All of these conclusions are in agreement with those of the present study.

DIAGNOSES OF MONOPHYLETIC GROUPS OF IGUANINES

In this section I provide discussions of the monophyletic groups of iguanines at and above the level of the basic taxa used in this study (traditional genera). For each taxon I include: (1) the type on which the taxon is based, (2) the etymology of the name, (3) a phylogenetic definition (de Queiroz, 1987; Gauthier et al., 1988), (4) the current distribution, (5) a diagnosis consisting of hypothesized synapomorphies, (6) fossil records, and (7) various comments. Synonyms are not provided; those of the basic taxa can be found in Etheridge (1982).

Iguaninae Bell 1825

Type genus: *Iguana* Laurenti 1768.

Etymology: Modification of *Iguana*, the name of its type genus.

Definition: The most recent common ancestor of *Brachylophus*, *Dipsosaurus*, and Iguanini, and all of its descendants.

Distribution: Southwestern United States southward through México, Central America, and northern South America to southern Brazil and Paraguay; the West Indies; the Galápagos Islands; Iles Wallis; and the Fiji and Tonga island groups.

Diagnosis: Iguanines are moderate to large iguanians that can be distinguished from other iguanians by the following synapomorphies:
 1. Vertebrae in part of caudal sequence bear two pairs of transverse processes (Etheridge, 1967).
 2. Transverse colic folds or valves present (Iverson, 1980, 1982).
 3. Crowns of posterior marginal teeth laterally compressed, anteroposteriorly flared, often with four or more cusps (Etheridge, 1964a).
 4. Supratemporal lies primarily on posteromedial surface of supratemporal process of parietal.
 5. Herbivorous (H. M. Smith, 1946; Iverson, 1982).

Fossil record: The diagnosis and description of iguanines presented here enable me to reject the possible iguanine relationships of certain fossil taxa. In their description of

Paradipsosaurus mexicanus, Fries et al. (1955:15) stated that this animal "would appear to approach more closely to the northern crested lizard *Dipsosaurus* than to any of the other iguanids that presently live in México and the southwestern United States." However, the similarities they cite (broad, flat parietal table elevated well above level of supratemporal arch; unrestricted supratemporal fossa; deep, broad snout without pronounced nasolachrymal ridges; forward opening nares), provide no evidence for a close relationship to *Dipsosaurus*, since they are all plesiomorphic for Iguania. Of the five diagnostic iguanine synapomorphies identified in this study, only the morphology of the tooth crowns can be assessed in *Paradipsosaurus*. Unlike the teeth of iguanines, those of *Paradipsosaurus* are said to be a little dilated and noncuspidate (Fries et al., 1955). Furthermore, while all postembryonic iguanines and various other iguanids have a relatively small splenial and have the dentary portion of Meckel's groove closed and fused, both derived features within Iguania, the splenial of *Paradipsosaurus* is relatively large and Meckel's groove is open (Estes, 1983). Therefore, although *Paradipsosaurus* and *Dipsosaurus* share the derived condition of having the parietal foramen located within the frontal bone, this similarity is convergent, since *Paradipsosaurus* is not an iguanine. Estes (1983) reached similar conclusions concerning the relationships of this fossil.

Gilmore (1928) described *Parasauromalus olseni* based on a fragment of a right dentary from the Eocene of Wyoming. Although he did not specifically propose that it was related to the iguanine *Sauromalus*, Gilmore considered the teeth of the fossil to resemble those of *Sauromalus ater* most closely, made his comparisons with this species only, and named the fossil as if to suggest a close relationship with *Sauromalus* (*para* means near). If new material has been correctly referred to *Parasauromalus* (Estes, 1983), then this taxon is not an iguanine and therefore cannot be closely related to *Sauromalus*. Contrary to Gilmore's (1928) statements, the tooth crowns of *Parasauromalus* are not particularly similar to those of *Sauromalus*. They are only slightly flared and tricuspid (Estes, 1983), while those of *Sauromalus* are strongly flared and polycuspate. The supratemporal of *Parasauromalus* lies on the lateral surface of the supratemporal process of the parietal (figured by Estes, 1983), whereas the supratemporal of iguanines lies in a derived position on the medial surface. The splenial of *Parasauromalus* is relatively large and the Meckelian groove closed but unfused (Estes, 1983), primitive iguanian characters not retained by any iguanine.

The oldest fossils referred to Iguaninae for which this reference cannot be rejected are Lower Miocene in age: *Tetralophosaurus* (Olson, 1937), a fragment of a lower jaw from Nebraska referred to *Dipsosaurus* by Estes (1983); a fragment of a lower jaw and a sacral vertebra from Florida (Estes, 1963); and another fragment of a lower jaw from Texas, referred to either *Ctenosaura* or *Sauromalus* by Stevens (1977). Because of their fragmentary nature, these specimens are not definitely referable to Iguaninae on the basis of synapomorphies. The oldest fossil that is clearly iguanine is a nearly complete skull from the Pliocene of southern California (Norell, 1983). These and other fossil records are given under the least inclusive taxon to which they belong or are most closely related.

Comments: Three of the five iguanine synapomorphies are presumably part of a single "adaptive syndrome." Both the iguanine dentition (Hotton, 1955) and colic valves (Iverson, 1980, 1982) are thought to be adaptations for a third iguanine character, herbivory. However, because this correlation of form and function does not extend to all herbivorous lizards, dentition, diet, and colic anatomy are here treated as separate characters.

Although Iguaninae was first used by Cope (1886), Bell (1825) is credited with authorship under the principle of coordination (Article 36, third edition of the *International Code of Zoological Nomenclature*). The content of Iguaninae as defined here differs from that of Cope's (1886) Iguaninae in that the former includes *Dipsosaurus* and *Sauromalus* while the latter does not. Iguaninae as defined here is identical in content to an unnamed subset of Cope's (1900) more inclusive Iguaninae and to Etheridge's (1964a, 1982) informal "iguanines."

In addition to the diagnostic iguanine characters given above, acceptance of the phylogenetic relationships proposed in this paper requires that the reduction or loss of the ventral process of the squamosal (character 18-A) be interpreted as an iguanine synapomorphy that has subsequently reversed in *Amblyrhynchus* and *Iguana*.

In order to facilitate diagnosis of the monophyletic subgroups of iguanines, I have reconstructed a hypothetical ancestral iguanine. This hypothetical ancestor has the derived characters of iguanines as a whole but lacks the derived characters of its monophyletic subgroups. The reason for constructing a hypothetical ancestor is that my diagnoses for the monophyletic subgroups of iguanines consist exclusively of synapomorphies, while it may also be useful to know what primitive features are retained by members of particular monophyletic subgroups. Members of any monophyletic subgroup of iguanines possess the condition found in the hypothetical ancestor unless an alternative state of the same character is listed as a diagnostic synapomorphy either of the taxon in question or of a larger monophyletic taxon of iguanines within which the taxon in question is included. It should be kept in mind that the presence of a primitive character properly indicates only that the specimen possessing it does not belong to the taxon diagnosed by the derived alternative condition. It does not preclude the possibility that the specimen in question, perhaps some newly discovered fossil, is not most closely related to the taxon diagnosed by the derived condition.

The hypothetical ancestral iguanine is thought to have possessed the following morphological features (numbers and letters correspond with those in the list of systematic characters):

1-A. Ventral surface of premaxilla bears large posterolateral processes.

2-A. Posteroventral crests of premaxilla small, not continuing up sides of incisive process and not pierced by foramina for maxillary arteries.

3-A. Anterior surface of premaxilla broadly convex.

4-A. Nasal process of premaxilla slopes posteriorly.

5-A. Nasal process of premaxilla exposed broadly between nasals.

6-A. Nasal capsule of moderate size, nasals relatively small.

7-A. Lacrimal contacts palatine, and prefrontal fails to contact jugal behind lacrimal foramen.

8-A. Frontal longer than wide.

9-A. Paired openings near frontonasal suture small or absent.

10-A. Cristae cranii of frontal form a smooth, continuous curve from frontal to prefrontal.

11-A. Frontal cristae medial to cristae cranii absent or weakly developed.

12-A. Dorsal borders of orbits form a more or less smooth curve.

13-A. Parietal foramen lies on frontoparietal suture.

14-A. Supratemporal extends anteriorly more than halfway across posterior temporal fossa.

15-A. Lateral surfaces of maxillae relatively flat or concave below supralabial foramina.

16-A. Premaxillary process of maxilla not curving dorsally; maxillary and premaxillary teeth lie in the same plane.

17-A. Lacrimal relatively large.

18-B. Ventral process of squamosal reduced or absent.

19-A. Squamosal does not abut against tympanic crest of quadrate.

20-A. Septomaxilla without pronounced longitudinal crest on anterolateral surface.

21-A. Palatine without high crest on dorsomedial surface.

22-A. Large posterolateral process of palatine behind infraorbital foramen present.

23-A. Posterolateral process of palatine behind infraorbital foramen fails to contact jugal. Contact of this process with the jugal may be a synapomorphy of all iguanines that has been lost secondarily in *Dipsosaurus*.

24-A. Infraorbital foramen located on lateral or posterolateral edge of palatine.

25-A. Medial borders of pterygoids relatively straight anterior to pterygoid notch, pyriform recess narrows gradually anteriorly. Sharply curved medial pterygoid borders and a pyriform recess that narrows abruptly may be a synapomorphy of all iguanines that has been secondarily lost in *Brachylophus*.

26-A. Ectopterygoid fails to contact palatine at posteromedial corner of suborbital fossa.

27-A. Long parasphenoid rostrum.

28-A. Cristae ventrolaterales of parabasisphenoid strongly constricted behind basipterygoid processes.

29-A. Posterolateral processes of parabasisphenoid large, extending far up anterior edges of lateral processes of basioccipital.

30-A. Laterally directed pointed process of cristae interfenestralis absent.

31-A. Stapes relatively thin.

32-A. Dorsal edges of dentary and surangular on either side of coronoid eminence approximately equal in height.

33-A. Splenial relatively large.

34-35-A. Anterior inferior alveolar foramen lies between splenial and dentary; coronoid may or may not contribute to its posterior margin.

36-A. Labial process of coronoid present but relatively small.

37-A. Angular extends far up lateral surface of mandible and is easily visible in lateral view.

38-A. Angular wide posteriorly.

39-A. Surangular does not extend anteriorly to last dentary tooth on labial surface of mandible.

40-A. Dome-shaped portion of surangular visible below coronoid on lingual surface of mandible.

41-A. Angular process of prearticular increases substantially in relative size during postembryonic ontogeny, becoming a prominent structure in adults.

42-A. Outline of retroarticular process triangular rather than quadrangular in all postembryonic developmental stages.

43-44-B. Mode of seven premaxillary teeth.

45-A. Lateral cusps of premaxillary teeth small or absent.

46-A. Posterior marginal teeth tricuspid. The presence of a fourth cusp may be a synapomorphy of all iguanines, with secondary loss in *Amblyrhynchus* and in some *Brachylophus* and *Ctenosaura*. Alternatively, the ancestral iguanine may have been polymorphic for the presence of a fourth cusp (again with secondary loss in *Amblyrhynchus* and some *Ctenosaura*).

47-A. Individual lateral cusps of tricuspid marginal teeth much smaller than apical cusp.

48-A. Entire pterygoid tooth row lies close to ventromedial edge of pterygoid.

49-A. Pterygoid tooth patch consists of a single row of teeth throughout postembryonic ontogeny.

50-A. Pterygoid tooth patch extends anteriorly beyond level of posterior edge of suborbital fenestra.

51-A. Pterygoid teeth present.

52-53-B. Second ceratobranchials from two-thirds length to slightly longer than first ceratobranchials.

54-A. Second ceratobranchials in medial contact for most or all their lengths.

55-A. Neural spines of presacral vertebrae tall, more than 50% of total vertebral height.

56-A. Zygosphenes connected to prezygapophyses by continuous arc of bone.

57-A. Posterolateral processes present on pleurapophyses of second sacral vertebra.

58-A. Foramina present in ventral surface of pleurapophyses of second sacral vertebra.

59-A. More than 40 caudal vertebrae.

60-A. Caudal autotomy septa present. The polarity of this character is questionable.

61-A. Autotomic caudal series (or series of caudal vertebrae with paired transverse processes) begins at or before 10th caudal vertebra. The polarity of this character is questionable.

62-A. Dorsal midsagittal fins of caudal vertebrae anterior to neural spines relatively large and present well beyond anterior third of caudal sequence.

63-A or -B. Postxiphisternal inscriptional ribs do not form continuous chevrons, or anteriormost pairs do only variably.

64-A. Suprascapulae oriented primarily vertically and form a continuous arc with the scapulocoracoids.

65-A. Scapular fenestrae present and large.

66-A. Posterior coracoid fenestrae absent.

67-A. Clavicles wide, with prominent lateral shelves.

68-A. Posterior process of interclavicle extends well beyond lateral corners of sternum.

69-A. Interclavicle arrow-shaped, lateral processes forming angles of less than 75° with posterior process.

70-A. Sternal fontanelle present and of moderate size.

71-A. Sternum diamond-shaped, xiphisternal rods attach close to midline.

72-A. Pelvic girdle relatively long and narrow.

73-A. Large anterior iliac process.

74-A. Cephalic osteoderms absent.

75-A. Heart lies entirely anterior to transverse axillary plane.

76-A. Subclavian arteries covered ventrally by posterior end of *M. rectus capitis anterior*.

77-A. Right and left systemic arches unite to form dorsal aorta above heart.

78-A. Coeliac artery arises from dorsal aorta anterior to and separate from mesenteric arteries.

79-A. Colic wall with one or more transverse valves.

80-A. All colic valves semilunar. The polarity of this character is questionable.

81-A. Median azygous rostral scale present.

82-A. Snout scales small and numerous, approximately same size as those of supraorbital and temporal regions.

83-A. Dorsal head scales flat or only slightly convex.

84-B. Superciliary scales moderately elongate and partially overlapping. It is also possible that the ancestral iguanine had elongate and strongly overlapping superciliaries.

85-A or -B. Subocular scales subequal in size, or one or two moderately elongate.

86-A. Anterior auricular scales small or only slightly enlarged.

87-A. Gular fold well developed.

88-A. Dewlap small or absent. The polarity of this character is questionable.

89-A. Gular crest of enlarged scales absent.

90-A. Middorsal scale row present.

91-A. Pedal subdigital scales asymmetrical, anterior keels larger than posterior ones.

92-A. Pedal subdigital scales lack greatly enlarged anterior keels fused at their bases to form combs.

93-A. Toes unwebbed.

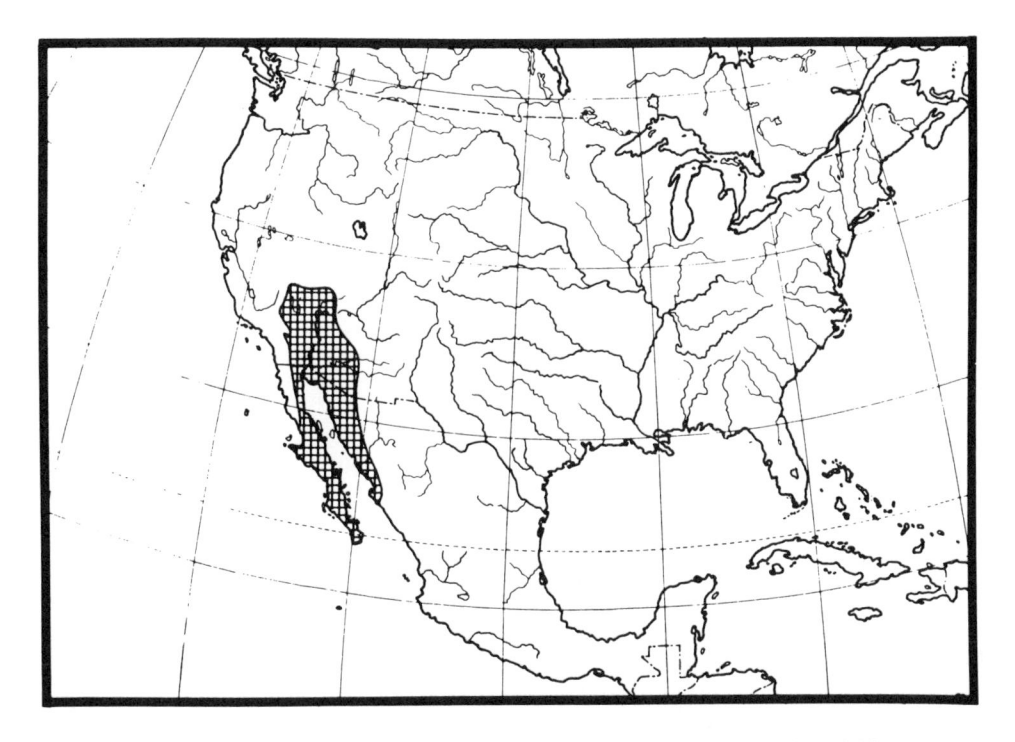

FIG. 52. Geographic distribution of *Dipsosaurus* (modified from Stebbins, 1966).

94-A. Caudal scales in adjacent verticils approximately equal in size, smooth or keeled but not spinous.

95-A. Body laterally compressed or roughly cylindrical.

Dipsosaurus Hallowell 1854

Type species (by monotypy): *Crotaphytus dorsalis* Baird and Girard 1852.

Etymology: (Greek) *Dipsa*, thirst(y), + *sauros*, lizard. *Dipsosaurus* was first known from the "Colorado Desert" of western North America, as Hallowell (1854:92) described it "a country without water."

Definition: The most recent common ancestor of the populations of Recent *Dipsosaurus dorsalis* and all of its descendants.

Distribution: Deserts of the southwestern United States in southeastern California, southern Nevada, southwestern Utah, and western Arizona, southward into México through western Sonora and northwestern Sinaloa and into Baja California to its southern end, including various islands in the Gulf of California (Fig. 52).

Diagnosis: Members of this taxon can be distinguished from other iguanines by the following synapomorphies (here and afterwards the parenthetical numbers and letters correspond with those in the list of systematic characters):

1. Large, paired openings at or near frontonasal suture present (9-B).

2. Parietal foramen located entirely within frontal bone (13-C). This character occurs also in *Cyclura carinata* and variably in some *Ctenosaura, Sauromalus,* and other *Cyclura.*

3. Lateral process of palatine behind infraorbital foramen small or absent (22-B).

4. Medial borders of pterygoids curve sharply toward midline anterior to pterygoid notch; pyriform recess narrows abruptly (25-B). This character occurs in all other iguanines except *Brachylophus* and may thus be a synapomorphy of Iguaninae that has reversed in *Brachylophus.*

5. Lateral pointed processes on cristae interfenestralis present (30-B).

6. Posterior ends of lateral and medial crests of retroarticular process diverge ontogenetically, so that outline of retroarticular process is quadrangular in large specimens (42-B).

7. Crowns of posterior marginal teeth with four cusps (46-B). An increase in tooth cuspation characterizes all other iguanines except *Amblyrhynchus* and some *Brachylophus* and *Ctenosaura*; therefore this character may be a synapomorphy of a more inclusive group that has reversed in certain taxa.

8. Pterygoid teeth usually absent (50-B, 51-B). This character also occurs in *Conolophus*. When present, the pterygoid teeth of *Dipsosaurus* lie along the medial edge of the pterygoid, while those of *Conolophus* lie more laterally, supporting the conclusion that the absence of pterygoid teeth in these two taxa is convergent.

9. Colon with one or more circular valves (80-B). This condition occurs also in all other iguanines except *Brachylophus* and may be a synapomorphy of a more inclusive group.

10. Superciliary scales greatly elongate and strongly overlapping (84-C). The derived status of this character is questionable.

11. One subocular scale much longer than others (85-C). The derived status of this character is questionable.

Fossil record: Olson (1937) described *Tetralophosaurus minutus* based on a fragment of a lower jaw from Lower Miocene deposits in Nebraska. The specimen was referred to *Dipsosaurus* by Estes (1983), who stated that it was indistinguishable from *D. dorsalis*, but this conclusion is based on overall similarity. Almost complete skulls and dentaries from the Pliocene of southern California have been referred to *Dipsosaurus* by Norell (1983).

Comments: Failure of the lateral palatine process to contact the jugal behind the infraorbital foramen (character 23) suggests that *Dipsosaurus* is the sister group of all other iguanines. However, the gently curving medial pterygoid borders and wide pyriform recess of *Brachylophus* (character 25) suggest that this taxon, rather than *Dipsosaurus*, is the sister group of all other iguanines. The weaker tendency of *Brachylophus* to develop fourth cusps on the posterior marginal teeth might be taken as further evidence in favor of the latter hypothesis, but the character is variable in *Brachylophus* and has reversed several other times within iguanines. At least three other characters might be used to support one or the other of these alternative hypotheses, but these characters must be used with caution because their polarities are unclear. These are: (1) the lack of a notch separating zygosphenes from prezygapophyses in *Dipsosaurus* (character 56); (2) the absence of circular colic valves in *Brachylophus* (character 80); and (3) the low number of colic valves in *Dipsosaurus* (Iverson, 1982). Camp (1923) noted another character in which all iguanines except *Dipsosaurus* share what appears to be a derived condition (*Conolophus* was not examined): a high degree of separation of the *M. mylohyoideus anterior superficialis*. Because of this contradictory information, I have chosen to leave the relationships among *Dipsosaurus*, *Brachylophus*, and the monophyletic group consisting of the remaining iguanines (Iguanini) unresolved. I am not aware of any characters suggesting that *Dipsosaurus* and *Brachylophus* are sister taxa.

Brachylophus Wagler 1830

Type species (by monotypy): *Iguana fasciata* Brongniart 1800.

Etymology: (Greek) *Brachys*, short, + *lophos*, a crest. The name presumably refers to the relatively short scales of the dorsal crest in *B. fasciatus*, the type species.

Definition: The most recent common ancestor of *B. fasciatus* and *B. vitiensis* and all of its descendants.

Distribution: Numerous islands in the Fiji Islands group, Tongatapu in the Tonga Islands group, and Iles Wallis northeast of Fiji, all in the southwestern Pacific Ocean (Fig. 53).

Diagnosis: Members of this taxon can be distinguished from other iguanines by the following synapomorphies:
　1. Lateral process of palatine behind infraorbital foramen contacts jugal (23-B). This character occurs in all iguanines except *Dipsosaurus* and some specimens of *Sauromalus*, and may be a synapomorphy of a more inclusive group.
　2. Infraorbital foramen located entirely within palatine bone, may or may not be connected to lateral edge of palatine by suture (24-B). This character also occurs in some *Amblyrhynchus*, some *Ctenosaura*, and some *Sauromalus*, in which it is interpreted as convergent.
　3. Anterior inferior alveolar foramen located entirely within dentary (34-35-B). This character occurs only in *Brachylophus* within Iguaninae, but does not occur in all specimens.

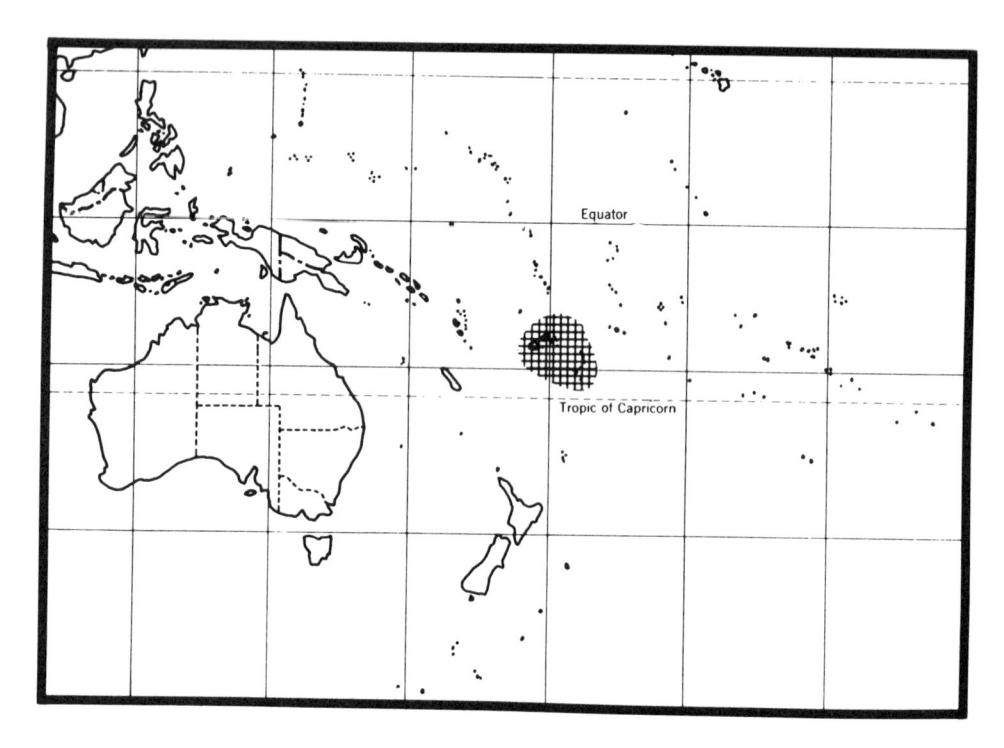

FIG. 53. Geographic distribution of *Brachylophus* (from Gibbons, 1981; Etheridge, 1982).

4. Labial process of coronoid moderately large (36-B). The enlarged labial coronoid process of *Amblyrhynchus* and *Conolophus* is interpreted as convergent.

5. Second ceratobranchials much longer than first ceratobranchials (52-53-C). The long second ceratobranchials of *Iguana iguana* are interpreted as convergent.

6. Zygosphenes separated from prezygapophyses by a deep notch (56-B). This character occurs in all iguanines except *Dipsosaurus,* and may be a synapomorphy of a more inclusive group.

7. Caudal autotomy septa absent (60-B). Although the outgroup evidence is equivocal, I have assumed that the presence of caudal autotomy, and the intravertebral septa that facilitate it, are primitive for iguanines. The absence of caudal autotomy septa in *Amblyrhynchus* and *Conolophus* on the one hand and in *Iguana delicatissima* on the other are interpreted as convergent.

8. Midsagittal processes on dorsal surfaces of caudal centra anterior to neural spine relatively small and confined to anterior fifth of caudal sequence (62-B). This character also occurs in *Iguana,* in which it is interpreted as convergent.

9. Anterior postxiphisternal inscriptional ribs enlarged and members of at least one pair united midventrally to form continuous chevrons (63-C). Midventrally continuous chevrons formed by the first pair of postxiphisternal inscriptional ribs occur in various other iguanines but not invariably within species, as in *Brachylophus*. Unlike other

iguanines, *Brachylophus* also exhibits enlargement of the second and third postxiphisternal inscriptional ribs, which may also unite to form continuous chevrons.

10. Large dewlap present (88-B). The two species of *Brachylophus* differ in that a large dewlap is present in both sexes of *B. vitiensis* but only in male *B. fasciatus* (Gibbons, 1981). The polarity of this character is uncertain. If presence of a large dewlap is derived, then the phylogenetic relationships proposed here require that it has evolved convergently in *Iguana* and in some species of *Ctenosaura*.

In addition, the following derived character occurs in some *Brachylophus*:
Posterior marginal teeth with a fourth cusp (46-B). This character occurs in all other iguanines except *Amblyrhynchus* and some *Ctenosaura*; it may thus be a synapomorphy of a more inclusive group, perhaps of all iguanines.

Fossil record: Bones thought to be remains of *Brachylophus* are known from archaeological sites on Tongatapu and Lifuka in the Tonga Islands group (approximately 2000 years before present). If correctly referred, these bones indicate that *Brachylophus* once reached much larger sizes than they do today (Etheridge, pers. comm.; Pregill, pers. comm.).

Comments: Gibbons (1981) discusses the authorship of *Brachylophus*, crediting the name to Wagler (1830), since Cuvier (1829) had used the informal apellation *les Brachylophes*. The relationships of *Brachylophus* to *Dipsosaurus* and other iguanines are discussed in the comments on *Dipsosaurus*, above.

Iguanini Bell 1825

Type genus: Iguana Laurenti 1768.

Etymology: Modification of *Iguana*, the name of its type genus.

Definition: The most recent common ancestor of *Ctenosaura, Sauromalus*, Amblyrhynchina, and Iguanina, and all of its descendants.

Distribution: Southwestern United States southward through México, Central America, and northern South America to southern Brazil and Paraguay, the West Indies, and the Galápagos Islands.

Diagnosis: Members of this taxon can be distinguished from other iguanines (*Brachylophus* and *Dipsosaurus*) by the following synapomorphies:
1. Lateral process of palatine contacts jugal behind infraorbital foramen (23-B). This character does not occur in some *Sauromalus*, where it is interpreted as a reversal. It does occur in *Brachylophus* and may thus be a synapomorphy of a more inclusive group.

2. Medial borders of pterygoids curve sharply toward midline anterior to pterygoid notch; pyriform recess narrows abruptly (25-B). This character occurs also in *Dipsosaurus* and may be a synapomorphy of a more inclusive group.

3. Crowns of posterior marginal teeth with four or more cusps (46-B,-C, or-D). This character occurs also in *Dipsosaurus* and some *Brachylophus*, and may be a synapomorphy of all iguanines. It has reversed in *Amblyrhynchus* and some *Ctenosaura*.

4. Posterior portion of pterygoid tooth patch displaced laterally away from medial border of pterygoid (48-B). Pterygoid teeth are absent in most *Conolophus*, but when present they lie away from the medial pterygoid border. This character develops during postembryonic ontogeny and is not always evident in small specimens.

5. Zygosphenes separated from prezygapophyses by a deep notch (56-B). This character occurs also in *Brachylophus* and may be a synapomorphy of a more inclusive group.

6. Sequence of autotomic caudal vertebrae or that of vertebrae with two pairs of transverse processes begins at or behind 10th caudal vertebra (61-B). The polarity of this character is questionable.

7. Posterior coracoid fenestra usually present (65-B). This character exhibits some variation within basic taxa.

8. Right and left systemic arches unite to form dorsal aorta posterior to heart (77-B).

9. One or more circular colic valves present (80-B). This character occurs also in *Dipsosaurus* and may be a synapomorphy of a more inclusive group.

Fossil record: The earliest fossils that are clearly referable to Iguanini are from the Pliocene of southern California. Among extant Iguanini these fossils appear to be most closely related to *Iguana* (Norell, 1983). Stevens (1977) considered a dentary fragment from the early Miocene of Texas to be either *Ctenosaura* or *Sauromalus*. If correctly referred, this would be the oldest record of Iguanini. These and other fossil records are given under the least inclusive taxa to which they belong or are most closely related.

Comments: Although this is the first use of Iguanini, Bell (1825) is credited with authorship under the principle of coordination (Article 36, third edition of the *International Code of Zoological Nomenclature*). Iguanini contains all the really large iguanines, and large body size may be an additional synapomorphy of this taxon. Some *Ctenosaura* are relatively small, but this probably represents a secondary reduction in size (see comments on *Ctenosaura*, below). Relationships among four recognizable monophyletic subgroups of Iguanini are uncertain and are discussed in greater detail in the comments on *Ctenosaura*, *Sauromalus*, Amblyrhynchina, Iguanina, and *Cyclura*.

Ctenosaura Wiegmann 1828

Type species (subsequent designation by Fitzinger 1843): *Ctenosaura cycluroides* Wiegmann 1828 = *Lacerta acanthura* G. Shaw 1802.

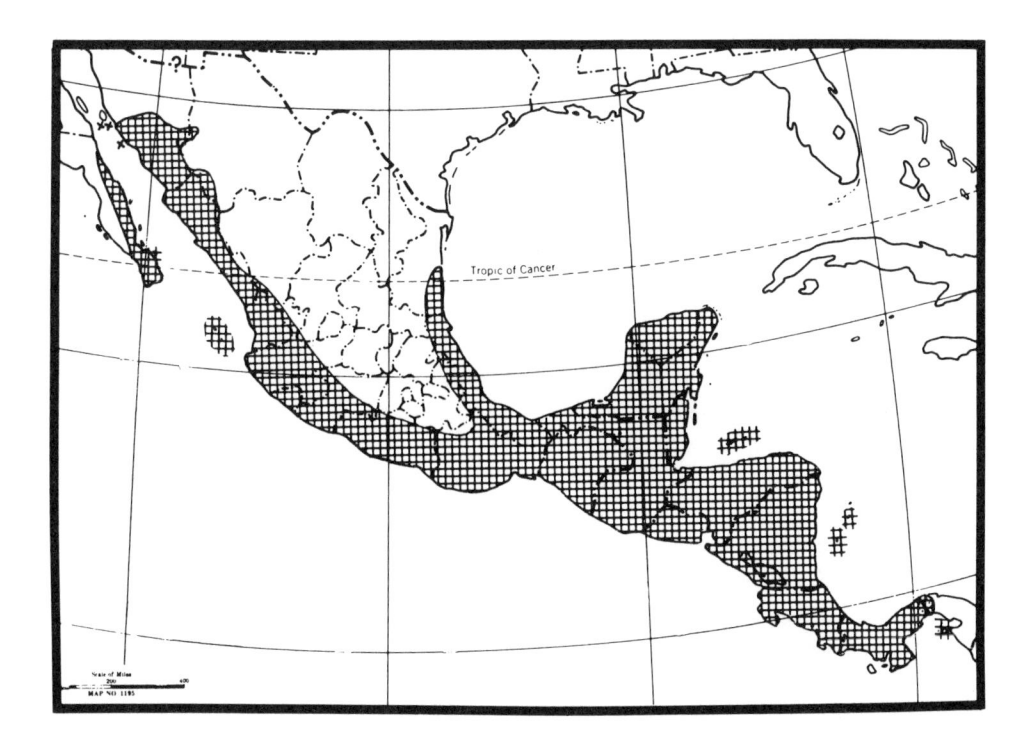

FIG. 54. Geographic distribution of *Ctenosaura* (from Peters and Donoso-Barros, 1970; H. M. Smith, 1972; Etheridge, 1982).

Etymology: (Greek) *Ktenos*, comb, + *sauros*, lizard, referring to the dorsal crest of enlarged scales.

Definition: The most recent common ancestor of the extant species of *Ctenosaura* (*acanthura, bakeri, clarki, defensor, hemilopha, palearis, pectinata, quinquecarinata,* and *similis*) and all of its descendants.

Distribution: Lowlands of México and Central America from southeastern Baja California and the middle of Sonora in western México and near the Tropic of Cancer in eastern México southward through most of Central America to central Panamá, as well as Isla de Providencia, Isla de San Andres, the Tres Marias Islands, and various offshore islands in the eastern Pacific, the western Caribbean, and the Sea of Cortez (Fig. 54).

Diagnosis: Members of this taxon can be distinguished from other iguanines by the following synapomorphies:
1. Premaxillary process of maxilla curves dorsally; premaxillary teeth set higher than maxillary teeth (16-B). This character is not present in small specimens.

2. Posterolateral processes of parabasisphenoid absent or relatively small (29-B).

3. Posterolateral processes on pleurapophyses of second sacral vertebra absent (57-B). This character also occurs in *Iguana* and most *Cyclura,* and may be a synapomorphy of a more inclusive group.

4. One subocular scale very long (85-C). The polarity of this character is questionable. An elongate subocular occurs also in *Dipsosaurus,* in which it is interpreted as convergent.

5. Tail bears whorls of enlarged, spinous scales (94-B). This character occurs also in most *Cyclura,* in which it is interpreted as convergent.

Other derived characters occur only in some *Ctenosaura* and may provide useful information concerning relationships within this taxon:

1. Prefrontal contacts jugal behind lacrimal foramen (7-B). This character also occurs in *Amblyrhynchus, Conolophus,* and some *Cyclura*; within *Ctenosaura,* prefrontal-jugal contact is characteristic only of *C. clarki* and may be a synapomorphy of that taxon.

2. Crista cranii forms step rather than smooth curve between frontal and prefrontal (10-B). This character also occurs in *Conolophus*; within *Ctenosaura* it occurs only in *C. defensor* and may be a synapomorphy of that taxon.

3. Parietal foramen located entirely within frontal (13-B). This character occurs also in *Dipsosaurus* and in some *Cyclura* and *Sauromalus*; within *Ctenosaura* it varies as much within species as among them, and it is therefore uninformative about relationships among these species.

4. Infraorbital foramen located entirely within palatine (24-B). This character also occurs in *Brachylophus* and in some *Amblyrhynchus* and *Sauromalus*; within *Ctenosaura* it varies as much within species as among them, and it is therefore uninformative about relationships among these species.

5. Surangular extends anteriorly well beyond coronoid apex and sometimes beyond posteriormost dentary tooth (39-B). This character occurs also in *Iguana* and *Cyclura*; its pattern of variation within *Ctenosaura* needs further study.

6. Crowns of posterior marginal teeth polycuspate (46-C). This character occurs also in *Iguana, Cyclura,* and *Sauromalus*; within *Ctenosaura* it occurs only in *C. defensor* and may be a synapomorphy of that taxon.

7. Crowns of posterior marginal teeth tricuspid (46-A). Within *Ctenosaura* this character, a presumed reversal, occurs in *C. bakeri* and *C. quinquecarinata.*

8. Posterior portion of pterygoid tooth patch doubles ontogenetically (49-B). This character, or a further modification of it, occurs also in *Iguana* and some *Cyclura.* Since members of the small species of both *Ctenosaura* and *Cyclura* do not exhibit ontogenetic doubling of the tooth row, and since small maximum size in these taxa is thought to be derived (see comments on Iguanini, above), it is likely that this character is a synapomorphy at a higher level and that failure to double the pterygoid tooth row is derived within *Ctenosaura.*

9. Fewer than 40 caudal vertebrae (59-B). This character also occurs in *Sauromalus*; within *Ctenosaura* it occurs in *C. clarki* and *C. defensor.*

10. Large dewlap (88-B). The polarity of this character is questionable. Large dewlaps occur also in *Brachylophus* and *Iguana*; within *Ctenosaura* they occur only in *C. palearis*.

Fossil record: The oldest fossils referred to *Ctenosaura* are from the Holocene of México (Langebartel, 1953; Ray, 1965; Estes, 1983). Stevens (1977) suggested that a fragment of a left dentary from the early Miocene of Texas was probably close to *Ctenosaura*.

Comments: Bailey (1928:7) claimed that "it is impossible to distinguish between the genus *Ctenosaura* and its near allies by means of skeletal characters." This is false. Osteological synapomorphies are identifiable not only in *Ctenosaura* but also in all of the other iguanine taxa that have traditionally been assigned the rank of genus. Even within *Ctenosaura*, monophyletic groups can be recognized on the basis of skeletal characters.

At least three characters suggest a close relationship among *Ctenosaura*, *Iguana*, and *Cyclura*: extension of the surangular well anterior to the coronoid apex (39-B); tendency of the pterygoid tooth row to double ontogenetically (49-B,-C); and absence of posterolateral processes on the pleurapophyses of the second sacral vertebrae (57-B). Nevertheless, I have left the relationships of *Ctenosaura* to other Iguanini unresolved because all three of these characters are ambiguous. The first is variably present in *Ctenosaura*, the third is variable in *Cyclura*, and the second is variable in both *Ctenosaura* and *Cyclura*. Thus, provided that the monophyly of each of these taxa is accepted, every one of these characters must involve homoplasy. If the homoplasy is interpreted as acquisition of the derived state of these characters in the most recent common ancestor of *Ctenosaura*, *Cyclura*, and *Iguana*, with subsequent reversal in certain taxa, then the close relationship among these three taxa might still be advocated. At present, however, the homoplasy can just as reasonably be interpreted as convergence, in which case the close relationship is not supported. I prefer to leave the relationships of *Ctenosaura* within Iguanini unresolved until additional evidence suggests that one of the alternative interpretations of homoplasy in the characters that vary within basic taxa is more plausible. The relationship between *Ctenosaura* and *Cyclura* is discussed further in the comments on *Cyclura*, below.

The species *bakeri*, *clarki*, *defensor*, *palearis*, and *quinquecarinata*, here included in *Ctenosaura*, are sometimes placed in a separate genus, *Enyaliosaurus*. Etheridge (1982) reviewed the history of the problem as follows:

> The most recent taxonomic revision and key for the genus *Ctenosaura* is that of Bailey (1928), but several important papers on individual species or groups of species have appeared subsequently. Bailey recognized 13 species, including those forms with a relatively small body size and a short, strongly spinose tail referred by some authors to *Enyaliosaurus*. Following Gray's (1845) description of *Enyaliosaurus* the name was seldom used until its revival by Smith and Taylor (1950: 75). In this work the species *clarki*, *defensor*, *erythromelas*, *palearis* and

quinquecarinata were allocated to *Enyaliosaurus*, but no justification was provided for the revival of the genus. Duellman (1965: 599), followed Smith and Taylor in recognizing the validity of *Enyaliosaurus*, placed *erythromelas* in the synonymy of *defensor*, provided a key to the species, and suggested that: "*Enyaliosaurus* doubtless is a derivative of *Ctenosaura*, all species of which are larger and have relatively longer tails and less well-developed spines than *Enyaliosaurus*." Meyer and Wilson (1973) referred *Ctenosaura bakeri* to *Enyaliosaurus*, but Wilson and Hahn (1973: 114-5) returned *bakeri* to *Ctenosaura*, commenting that: "John R. Meyer is currently studying the problems of the relationship of the species now grouped in *Enyaliosaurus* to those now grouped in *Ctenosaura*. He (pers. comm.) advised us that he considers the two genera inseparable, and that *bakeri* appears to be closely related to both *palearis* (now in *Enyaliosaurus*) and *similis* (now in *Ctenosaura*)." In addition, Ernest Williams of Harvard University has informed me (pers. comm.) that based on an unpublished study of the group by him and Clayton Ray, he does not believe the recognition of *Enyaliosaurus* is warranted. At the present time the problem of the relationships of *Ctenosaura* and *Enyaliosaurus* are under study by Diderot Gicca of the Florida State Museum. (Etheridge, 1982:9-10)

More recently, Gicca (1983) recognized the genus *Enyaliosaurus*.

Evidence for the monophyly of *Ctenosaura* in the broad sense of Bailey (1928) has been presented above. An evaluation of the monophyletic status of *Ctenosaura* in the narrow sense, and of *Enyaliosaurus*, required a phylogenetic analysis using the species of both as basic taxa. In this analysis, I have used primarily characters recognized by previous workers, in particular, Bailey (1928), Smith and Taylor (1950), and Ray and Williams (unpubl.). When possible, all characters were checked on specimens. My analysis is based on the following 19 characters representing a minimum of 23 phylogenetic transformations. The polarities of these characters were determined using Amblyrhynchina, Iguanina, *Sauromalus, Dipsosaurus,* and *Brachylophus* as outgroups. The character-state codes are as follows: 0, ancestral; 1, derived; 2, further derived; etc. Letter codes are used for characters whose polarities were considered undeterminable.

1. Maximum snout-vent length: (0) greater than 190 mm; (1) less than 190 mm. Maximum snout-vent lengths for the various taxa are as follows: *acanthura* = 215 mm (MCZ 16074, Bailey, 1928; 315 mm according to Ray and Williams, unpubl., but they include *pectinata* in *acanthura*); *bakeri* = 210 mm (USNM 25324, Bailey, 1928); *clarki* = 154 mm (UMMZ 112711, Duellman and Duellman, 1959); *defensor* = 155 mm (HM 3420, Bailey, 1928); *hemilopha* = approximately 400 mm (H. M. Smith, 1972; the largest specimen that Bailey [1928] presents data for is AMNH 2073 with a snout-vent length of 260 mm); *palearis* = 254 mm (CAS 69308, A. Bauer, pers. comm.); *pectinata* = 305 mm (MCZ 2726, Bailey, 1928); *quinquecarinata* = 169 mm (Hidalgo, 1980; Gicca, 1983); and *similis* = 489 mm (Fitch and Hackforth-Jones, 1983). A cutoff of 190 mm was chosen, partly because of an apparent gap and partly because all other species of Iguanini reach greater maximum snout-vent lengths than this.

2. Modal number of presacral vertebrae (Table 4): (0) 24; (1) 25.

3. Modal number of premaxillary teeth (Table 3): (0) seven; (1) five. Although *Ctenosaura defensor* is the only species with a mode of five premaxillary teeth (range 5-6), the occurrence of five premaxillary teeth in some specimens of *C. clarki* and *C. quinquecarinata*, but in no other *Ctenosaura*, suggests that these three species form a monophyletic group.

4. Anterior orbital region (Fig. 10): (A) lacrimal contacts palatine behind lacrimal foramen; (B) prefrontal contacts jugal behind lacrimal foramen.

5. Cristae cranii (Fig. 12): (0) form smooth curve from frontal to prefrontal; (1) frontal portions protrude anteriorly forming a step from frontal to prefrontal.

6. Parietal roof: (0) remains deeply notched posteriorly throughout ontogeny, so that the braincase is broadly exposed in dorsal view; (1) extends posteriorly as a flat shelf during postembryonic ontogeny, so that the braincase comes to be largely covered in dorsal view. This character is partially correlated with character 1, body size.

7. Ontogenetic convergence of lateral edges of parietal roof: (A) eventually meet posteriorly and form a midsagittal crest, giving the parietal roof a Y-shaped outline; (B) fail to meet, or meet but fail to form a midsagittal crest, giving the parietal roof a trapezoidal or triangular outline. This character is partially correlated with character 1, body size.

8-9. Crowns of posterior marginal teeth: (A0) with a maximum of four cusps; (B0) with a maximum of five or more cusps; (A1) with a maximum of three cusps.

10. Pendulous dewlap: (0) absent; (1) present but small; (2) present and large.

11. Parietal eye: (0) conspicuous externally; (1) external signs inconspicuous or absent. This character may also be manifested in a reduction in the parietal foramen in *C. defensor*, but my osteological sample of this taxon is small (N=1).

12. Dorsal crest scales I: (0) conform in color and pattern to adjacent body scales; adjacent crest scales similar in size; (1) unicolored and differing from body color; large, flap-like crest scales separated by one or more smaller scales.

13. Dorsal crest scales II: (0) high-keeled, large, and conspicuous, at least in neck region; (1) low-keeled to flat, inconspicuous throughout length of crest.

14. Middorsal scale row: (0) continuous from neck onto tail, or narrowly interrupted in sacral region; (1) broadly discontinuous in lumbosacral region.

15. Scales of anterodorsal surface of leg: (0) not enlarged or spinous; (1) enlarged and spinous on shank but not on thigh; (2) enlarged and spinous on both shank and thigh. An additional state could be recognized, since *C. clarki* and *C. quinquecarinata* have large anterodorsal thigh scales compared to those of most other *Ctenosaura*, but these scales are not as large as in *C. defensor*, and they are not spinous.

16. Subdigital scales at the base of pedal digit III: (0) with relatively small anterior keels or with moderately large anterior keels that are separate from those of adjacent scales; (1) with relatively large anterior keels fused at their bases to form a comb.

17. Tail: (0) strongly spinose proximally, but not distally, and always longer than body (snout-vent length/total length = 0.27-0.45), more than 30 caudal vertebrae; (1) tail

strongly spinose throughout its length and almost the same length as the body (snout-vent length/total length = 0.48-0.56), fewer than 30 caudal vertebrae.

18. Anterior (referring to first 10) whorls of strongly spinous caudal scales: (0) always separated by at least two rows of intercalary scales; (1) at least some separated by only one intercalary scale row, others by two or more; (2) none (or only the first) separated by two intercalary scale rows, but all separated by at least one; (3) intercalary scales of proximal whorls greatly reduced or absent.

19. Snout region: (0) not inflated, sloping gradually downward; (1) inflated anteriorly, sloping abruptly downward.

Height of the vertebral neural spines may also be a useful character, but I have chosen not to use it because I have no postcranial skeletons of *C. defensor* and *C. palearis*.

The distributions of these character states among basic taxa within *Ctenosaura* (*sensu lato*) and three near (Amblyrhynchina, Iguanina, *Sauromalus*) and two more distant (*Dipsosaurus*, *Brachylophus*) outgroups are given in Table 10. *Ctenosaura bakeri* from Isla de Utila and those from Isla de Roatán are scored separately because they differ in at least three of the characters used in this analysis. Only those from Utila, the type locality, are included in the analysis of relationships.

The phylogenetic relationships suggested by the characters in Table 10 (except character 19, the derived state of which occurs only in the Roatán population of *C. bakeri*) are diagrammed in Figure 55. Synapomorphies for the subterminal nodes and the basic taxa are given below. Characters whose polarities were initially undeterminable were placed on the cladogram after it was constructed using only those characters whose polarities were determinable using other iguanines as outgroups. Ignoring the Roatán ctenosaurs and intraspecific variation, these relationships require a total of 25 character transformations, three more than the minimum number required by the characters themselves (C-index = 0.88). C-indices for the individual characters are given in Table 10.

Node 1: *Ctenosaura* Wiegmann 1828

See above. The characters of the hypothetical ancestral *Ctenosaura* can be reconstructed by taking the first state of each of the 19 characters in the character list.

Node 2 (unnamed)

1. Parietal roof extends posteriorly over braincase during postembryonic ontogeny (6-1).

Ctenosaura acanthura

No synapomorphies identified.

Ctenosaura pectinata

No synapomorphies identified.

TABLE 10. Distributions of Character States of 19 Characters Among Basic Taxa Within *Ctenosaura* (in the broad sense) and Three Close and Two More Distant Outgroups

Taxon	Character																		
	1	2	3	4	5	6	7	8	9	10	11	12	13	14	15	16	17	18	19
acanthura	0	0	0	A	0	1	A	A	0	0	0	0	0	0	0	0	0	0	0
bakeri (Utila)	0	0	0[1]	-	-	0	B	A	1	1	0	1	0	0	1	0	0	0,1	0
bakeri (Roatán)	0	0	0	A	0	0	B	A	1	0	0	0	0	1	0	0	0,1	1
clarki	1	1	0	B	0	0	B	A	0	0	0	0	1	0,1	1	0	1	2	0
defensor	1	1	1	A	1	0	B	B	0	0	1	0	1	0,1	2	1	1	3	0
hemilopha	0	0	0	A	0	0	A	A	0	0	0	0	0	1	0	0	0	1	0
palearis	0	0	0	A	0	0	B	A	0	2	0	1	0	0	1	0	0	2	0
pectinata	0	0	0	A	0	1	A	A	0	0	0	0	0	0	0	0	0	0	0
quinquecarinata	1	1	0	A,B	0	0	B	A	1	0	0	0	0	1	0	0	2	0
similis	0	0	0	A	0	0	A	A	0	0	0	0	0	0	0	0	0	0	0
CI	1.0	1.0	1.0	1.0	1.0	1.0	1.0	1.0	0.5	1.0	1.0	1.0	1.0	1.0	1.0	1.0	1.0	0.6	1.0
Amblyrhynchina	0	0	0	B	0,1	0	A,B	A	0,1	0	0	0[2]	0	0	0	0	0[3]	N[3]	0
Iguanina	0	0	0[4]	A,B	0	0	A,B	B	0	1,2	0	0	0	0	0	0,1	0[3]	0,N[3]	0,1
Sauromalus	0	0	1[5]	A	0	0	B	B	0	0	0	N[6]	N[6]	1[6]	0[7]	0	0[3,8]	N[3]	0
Dipsosaurus	1	0	0	A	0	0	B	A	0	0	0	0	1	0	0	0	0[3]	N[3]	0
Brachylophus	0	0	0	A	0	0	A,B	A	0,1	0,1	0	0	0	0	0	0	0[3]	N[3]	0

Note: Character-state codes correspond with those used in the character list. A dash indicates the lack of data. Consistency indices (CI) for each character on the minimum-step cladogram for these characters (Fig. 55) are also given. The consistency indices were calculated ignoring intraspecific variation.
[1] 50% have seven and 50% have six (N = 2).
[2] Large, conical crest scales are separated by smaller ones in *Conolophus*.
[3] Not spinose in *Amblyrhynchus*, *Conolophus*, *Iguana*, *Sauromalus*, *Dipsosaurus*, and *Brachylophus*.
[4] Greater than seven in *Cyclura*.
[5] Some species have modes of four or six.
[6] Middorsal scale row entirely absent.
[7] In *S. hispidus* the entire leg has enlarged, spinous scales.
[8] Tail about same length as body but not spinose.

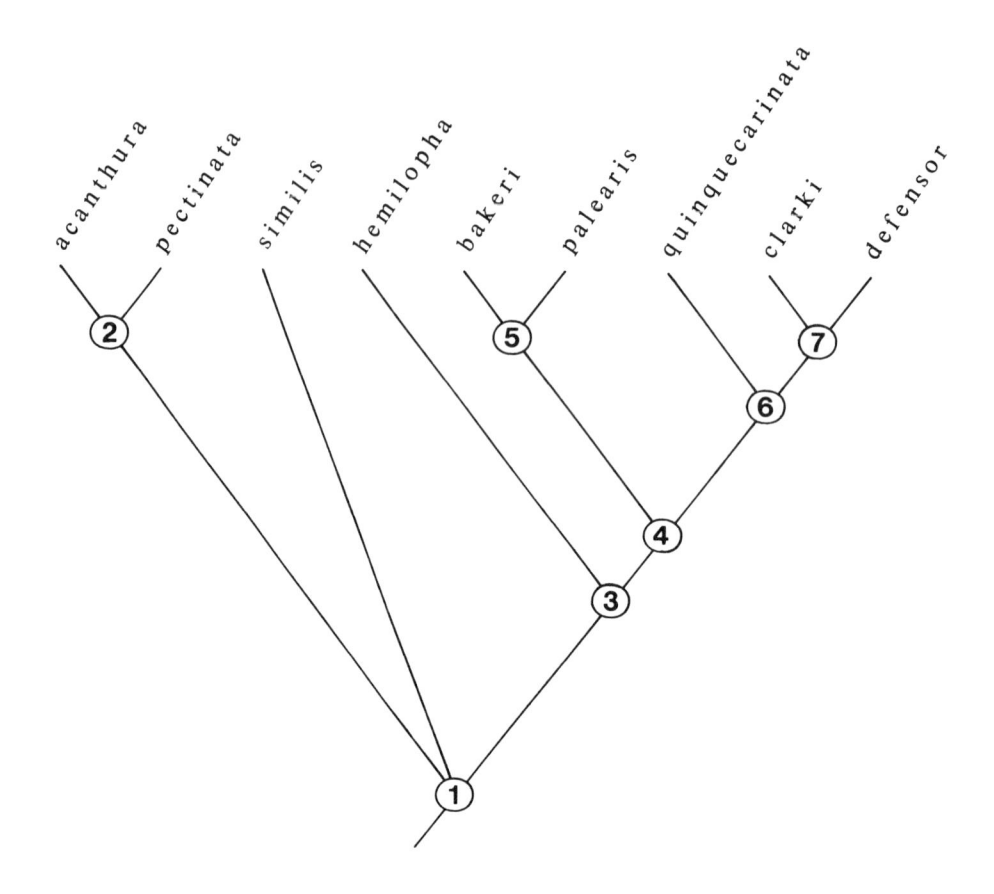

FIG. 55. Cladogram illustrating phylogenetic relationships within *Ctenosaura*. Synapomorphies for the numbered nodes and basic taxa are given in the text.

Node 3 (unnamed)
1. Some anterior whorls of strongly spinous caudal scales separated by one or fewer rows of smaller scales (18-1,-2, or-3). This character does not occur in all *C. bakeri*, a case most simply interpreted as a character reversal.

Ctenosaura hemilopha
1. Middorsal scale row broadly discontinuous in lumbosacral region (14-1). This character also occurs in some *C. clarki* and *C. defensor*, in which it is interpreted as convergent.

Node 4: *Enyaliosaurus* Gray 1845
1. Lateral edges of parietal roof fail to meet, or meet and fail to form a midsagittal crest; outline of parietal roof trapezoidal or triangular (7-B).
2. Scales on anterodorsal surface of shank enlarged and spinous (15-1 or-2).

Node 5 (unnamed)
1. Pendulous dewlap present (10-1 or-2).
2. Dorsal crest scales unicolored and differing in color from adjacent body scales; large, flap-like crest scales separated by one or more smaller scales (12-1).

Ctenosaura bakeri
1. Crowns of posterior marginal teeth with a maximum of three cusps (9-1). This character occurs also in *C. quinquecarinata*, where it is interpreted as convergent.

In some: anterior whorls of strongly spinous caudal scales always separated by at least two rows of smaller scales (18-0). This is interpreted as a case of character reversal.

Ctenosaura palearis
1. Large pendulous dewlap present (10-2).
2. All anterior whorls of strongly spinous caudal scales (except sometimes the first) separated by one or no intercalary scale rows (18-2 or-3). This character occurs also in *C. defensor*, *C. clarki*, and *C. quinquecarinata*, in which it is interpreted as convergent; alternatively, it may be a synapomorphy of a more inclusive group (*Enyaliosaurus*).

Node 6 (unnamed)
1. Maximum snout-vent length less than 190 mm (1-1).
2. Mode of 25 presacral vertebrae (2-1).
3. All anterior whorls (except sometimes the first) of strongly spinous caudal scales separated by one or no intercalary scale rows (18-2 or 3). This character occurs also in *C. palearis*, where it is interpreted as convergent; alternatively, it may be a synapomorphy of a more inclusive group.

The occurrence of five premaxillary teeth in at least some specimens, as well as enlargement of the scales on the anterodorsal surface of the thigh, may also be synapomorphies of this group.

Ctenosaura quinquecarinata
1. Crowns of posterior marginal teeth with a maximum of three cusps (9-1). This character occurs also in *C. bakeri*, in which it is interpreted as convergent.

In some: prefrontal contacts jugal behind lacrimal foramen (4-B). This character occurs also in *C. clarki*, in which it is interpreted as convergent.

Node 7 (unnamed)

1. Dorsal crest scales low-keeled to flat; inconspicuous throughout length of crest (13-1).

2. Tail strongly spinose throughout its length, and about same length as body (snout-vent/tail length = 0.48-0.56); fewer than 30 caudal vertebrae (17-1).

In some: middorsal scale row broadly discontinuous in lumbosacral region (14-1). This character occurs also in *C. hemilopha*.

Ctenosaura clarki

1. Prefrontal contacts jugal behind lacrimal foramen (4-B). This character occurs also in some *C. quinquecarinata*, in which it is interpreted as convergent.

Ctenosaura defensor

1. Mode of five premaxillary teeth (3-1).
2. Frontal portion of crista cranii projects anteriorly to form a step from frontal to prefrontal bones (5-1).
3. Crowns of posterior marginal teeth with a maximum of five or more cusps (8-B).
4. External signs of parietal eye inconspicuous or absent (11-1).
5. Scales on anterodorsal surface of thigh enlarged and spinous (15-2).
6. Anterior keels of subdigital scales at base of pedal digit III enlarged and fused at their bases to form a comb (16-1).
7. Proximal rows of smaller scales between whorls of enlarged, spinous caudal scales small or absent (18-3).

The results of the present analysis indicate that *Enyaliosaurus* (including *bakeri*, *palearis*, *quinquecarinata*, *clarki*, and *defensor*) is a monophyletic group, but that *Ctenosaura* in the narrow sense (*acanthura*, *pectinata*, *similis*, and *hemilopha*) is not. *Ctenosaura hemilopha* appears to have shared a more recent common ancestor with *Enyaliosaurus* than with the other *Ctenosaura* (in the narrow sense). However, the character that suggests an exclusive common ancestry for *hemilopha* and *Enyaliosaurus*, a reduction in the number of intercalary scale rows between the whorls of enlarged, spinous caudal scales, is problematical, in that it does not occur in all *bakeri*. Nevertheless, if *bakeri* and *palearis* are sister taxa, it is simpler to interpret the incongruence as a reversal in some *bakeri* rather than four separate acquisitions of the derived condition (1-in *hemilopha*, 2-in some *bakeri*, 3-in *palearis*, 4-in the common ancestor of *quinquecarinata*, *clarki*, and *defensor*). In any case, the monophyly of *Ctenosaura* in the narrow sense is doubtful even if this character is rejected, for there are no derived characters found in *acanthura*, *pectinata*, *similis*, and *hemilopha* that are not also found in the other taxa. Rather than the two being separate taxa, *Enyaliosaurus* appears to be a subgroup of a more inclusive *Ctenosaura*.

There currently exist several problems concerning species-level taxa within *Ctenosaura*. Smith and Taylor (1950) considered the specimens from the west coast of México assigned to *C. acanthura* by Bailey (1928) to be *C. pectinata*. Based on a conflict between the supposed geographic ranges of the species and the actual geographic distribution of specimens possessing the diagnostic characters of each species, Ray and Williams (unpubl.) considered *pectinata* to be a synonym of *acanthura*. The primary character used by Bailey (1928) to distinguish between these two taxa was whether the middorsal scale row was continuous (*pectinata*) or interrupted (*acanthura*) in the sacral region, but Hardy and McDiarmid (1969) claim that this character is variable within *pectinata* from western México. Nevertheless, synonymizing *pectinata* with *acanthura* on the basis of such data rests on an assumption that the two taxa are not broadly sympatric.

Stejneger (1901) described *Ctenosaura bakeri* from Utilla (Utila) Island, Honduras, and Bailey (1928) surmised that it may also occur on Bonacca (Guanaja) and Ruatan (Roatán) islands. Specimens collected subsequently on Roatán have been considered to be *C. bakeri* (Wilson and Hahn, 1973; Meyer and Wilson, 1973), but they differ from the Utila specimens in several ways (Table 10), including characters suggesting that they may not even be one another's closest relatives. The populations from the two islands are probably best considered separate species.

Sauromalus Duméril 1856

Type species (by monotypy): *Sauromalus ater* Duméril 1856.

Etymology: (Greek) *Sauros*, lizard, + *omalos*, flat.

Definition: The most recent common ancestor of the Recent species of *Sauromalus* (*ater, australis, hispidus, obesus, slevini,* and *varius*) and all of its descendants.

Distribution: Deserts of the southwestern United States in southeastern California, southern Utah and Nevada, and western and central Arizona, southward into México in western Sonora and eastern Baja California as well as various islands in the Gulf of California (Fig. 56).

Diagnosis: Members of this taxon can be distinguished from other iguanines by the following synapomorphies:

1. Parietal foramen located variably within frontal (13-B). This character occurs also in some populations of *Ctenosaura* and *Cyclura*; the parietal foramen is invariably located within the frontal in *Dipsosaurus* and *Cyclura carinata*.

2. Splenial relatively small (33-B).

3. Angular does not extend far up labial surface of dentary and is not visible, or is only barely visible in lateral view (37-B). This character also occurs in *Amblyrhynchus* and *Conolophus*; thus, it is either convergent or a synapomorphy of a more inclusive taxon.

4. Angular reduced and narrow posteriorly (38-B).

5. Modal number of premaxillary teeth fewer than seven (absolute range 3-7; range of modes for species 4-6) (43-44-A). This character also occurs in *Ctenosaura defensor*.

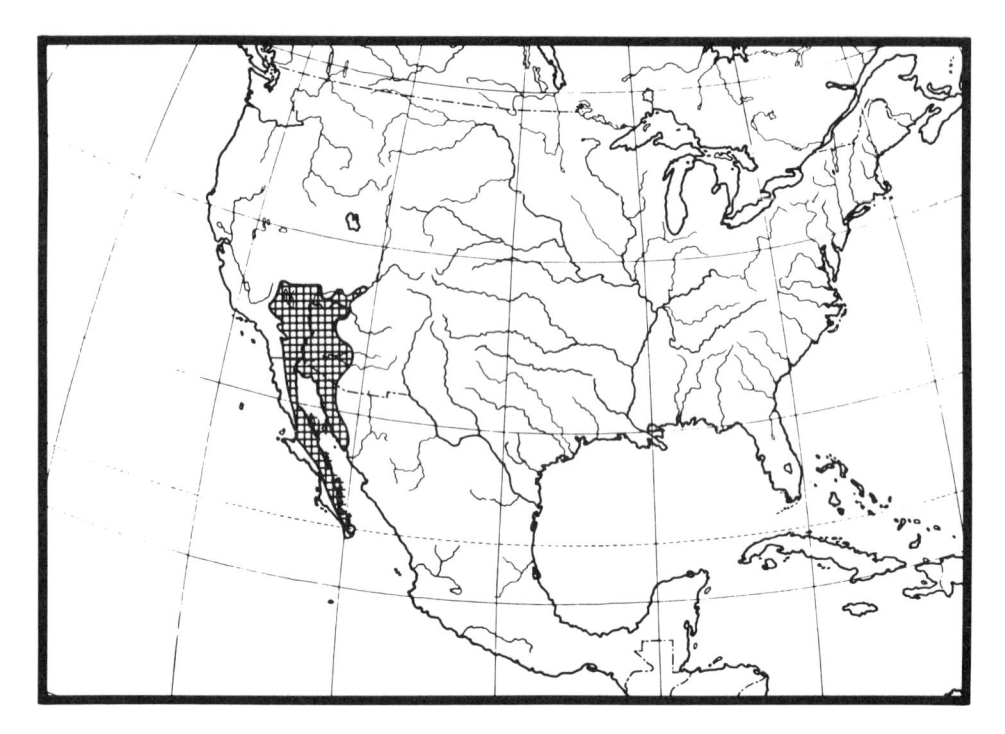

FIG. 56. Geographic distribution of *Sauromalus* (from C. E. Shaw, 1945; Gates, 1968; Etheridge, 1982).

6. Crowns of posterior marginal teeth with five or more cusps (46-C). This character occurs in *Cyclura* and *Iguana* and may thus be a synapomorphy of a more inclusive group. It also occurs in *Ctenosaura defensor*, in which it is interpreted as convergent.

7. Second ceratobranchials of hyoid apparatus short, often less than two-thirds the length of the first ceratobranchials (52-53-A). This character also occurs in *Amblyrhynchus* and *Conolophus,* in which it is either convergent or a synapomorphy of a more inclusive taxon.

8. Second ceratobranchials not in contact medially for most or all of their lengths (54-B). This character also occurs in *Amblyrhynchus*, in which it is interpreted as convergent.

9. Neural spines of presacral vertebrae short, less than 50% of total vertebral height (55-B).

10. Fewer than 40 caudal vertebrae (59-B). This character occurs also in *Ctenosaura clarki* and *C. defensor*, in which it is interpreted as convergent.

11. Postxiphisternal inscriptional ribs never form continuous midventral chevrons (63-A). The polarity of this character is questionable. It also occurs in *Dipsosaurus*, in which, if derived, it is interpreted as convergent.

12. Suprascapular cartilages situated primarily in a horizontal plane, and each forms an angle rather than a smooth curve with the scapula (64-B).

13. Scapular fenestrae small or absent (65-B). This character occurs also in *Amblyrhynchus*, in which it is interpreted as convergent.

14. Clavicles narrow, the lateral shelf small or absent (67-B).

15. Posterior process of interclavicle does not extend beyond lateral corners of sternum (68-B). This character also occurs in *Amblyrhynchus*, in which it is interpreted as convergent.

16. Lateral processes of interclavicle form angles of between 75° and 90° with posterior process, interclavicle roughly T-shaped (69-B). This character also occurs in *Amblyrhynchus*, in which it is interpreted as convergent.

17. Sternal fontanelle small or absent (70-B). This character occurs also in *Amblyrhynchus*, in which it is interpreted as convergent.

18. Sternum pentagonal; xiphisterna widely separated (71-B). This condition is approached in *Amblyrhynchus*.

19. Pelvic girdle short and broad (72-B).

20. Anterior iliac process small (73-B).

21. Heart extends posterior to transverse axillary plane (75-B).

22. Rostral scale divided by a median suture (81-B).

23. Superciliary scales quadrangular and non-overlapping (84-A). This character also occurs in *Amblyrhynchus*, in which it is interpreted as convergent.

24. Enlarged anterior auricular scales (85-B).

25. Middorsal scale row absent (89-B).

26. Anterior and posterior keels of subdigital scales approximately equal in size; subdigital scales roughly symmetrical with respect to long axis of toe (91-B).

27. Body strongly depressed (95-B).

Another possible synapomorphy of *Sauromalus* is the failure of the lateral edges of the parietal table to meet ontogenetically. This character occurs also in *Dipsosaurus* and in some *Brachylophus*, *Ctenosaura*, and *Cyclura*.

In addition, the following derived characters occur in some *Sauromalus*:

1. Lateral process of palatine behind infraorbital foramen fails to contact jugal (23-A). This reversal occurs variably within *ater*, *hispidus*, and *varius*.

2. Infraorbital foramen located entirely within the palatine (24-B). This character occurs also in *Brachylophus* and in some *Amblyrhynchus* and *Ctenosaura*. Within *Sauromalus* it is known only in *obesus*, and its occurrence is variable in this taxon.

3. Coeliac artery originates between mesenteric arteries (78-B). This character occurs also in *Iguana*. Its pattern of variation is poorly known, owing to small samples.

Fossil record: Estes (1983) summarized information on fossil *Sauromalus*, and additional material has been described subsequently by Norell (1986). The oldest fossils referred to this taxon are from the Pleistocene of California, Nevada, and Arizona. Stevens (1977) referred a fragment of a left dentary from the lower Miocene of Texas to either *Ctenosaura* or *Sauromalus*, but thought that it was probably closer to *Ctenosaura*.

Comments: Sauromalus has a large number of derived characters supporting its monophyly, many of which are unambiguous (i.e., they do not occur in any other iguanine). Although several of these characters, such as the height of the neural spines, the orientation of the suprascapulae, and the shape of the pelvic girdle, may be part of a single adaptive complex manifested externally in a depressed body form, I have treated them as separate synapomorphies. Because various combinations of the alternative states of these morphologies occur in certain noniguanine taxa, there is no reason to believe that they must always occur together.

Sauromalus shares a large number of derived characters with *Amblyrhynchus*, particularly in the shoulder girdle but not confined to this structure. Because *Amblyrhynchus* shares even more derived characters with *Conolophus*, and because *Conolophus* does not possess most of the derived characters shared by *Amblyrhynchus* and *Sauromalus*, I interpret the derived characters shared by *Amblyrhynchus* and *Sauromalus* as convergences. Perhaps this convergence results from similar functional demands placed on the shoulder girdle by the saxicolous habits of the animals in both taxa. The situation is complicated by the fact that all three taxa share two other derived characters: limited lateral exposure of the surangular (37-B), and relatively short second ceratobranchials (52-53-B). If one were to accept a sister-group relationship between the Galápagos iguanas and *Sauromalus* based on these characters, then the many characters shared by *Amblyrhynchus* and *Sauromalus*, but not *Conolophus*, could be interpreted as additional synapomorphies of this hypothesized clade that have reversed in *Conolophus*. However, because *Sauromalus* also shares a derived tooth morphology with *Cyclura* and *Iguana* that does not occur in the Galápagos iguanas, I see no compelling reason to accept a sister-group relationship between *Sauromalus* and the Galápagos iguanas. Furthermore, even if such a relationship were accepted, interpreting the derived characters shared by *Amblyrhynchus* and *Sauromalus* as convergent requires no more evolutionary changes than hypothesizing a single origin for the derived state of each character, with reversal in *Conolophus*.

The most recent revision of *Sauromalus* is that of C. E. Shaw (1945), although works of taxonomic significance have appeared subsequently (Cliff, 1958; Tanner and Avery, 1964; Soule and Sloan, 1966; Robinson, 1972, 1974). The boundaries, monophyly, and relationships of the species within *Sauromalus* need further study.

Amblyrhynchina, new taxon

Type genus: Amblyrhynchus Bell 1825.

Etymology: Modification of *Amblyrhynchus*, the name of its type genus.

Definition: The most recent common ancestor of the extant Galápagos iguanas, *Amblyrhynchus* and *Conolophus*, and all of its descendants.

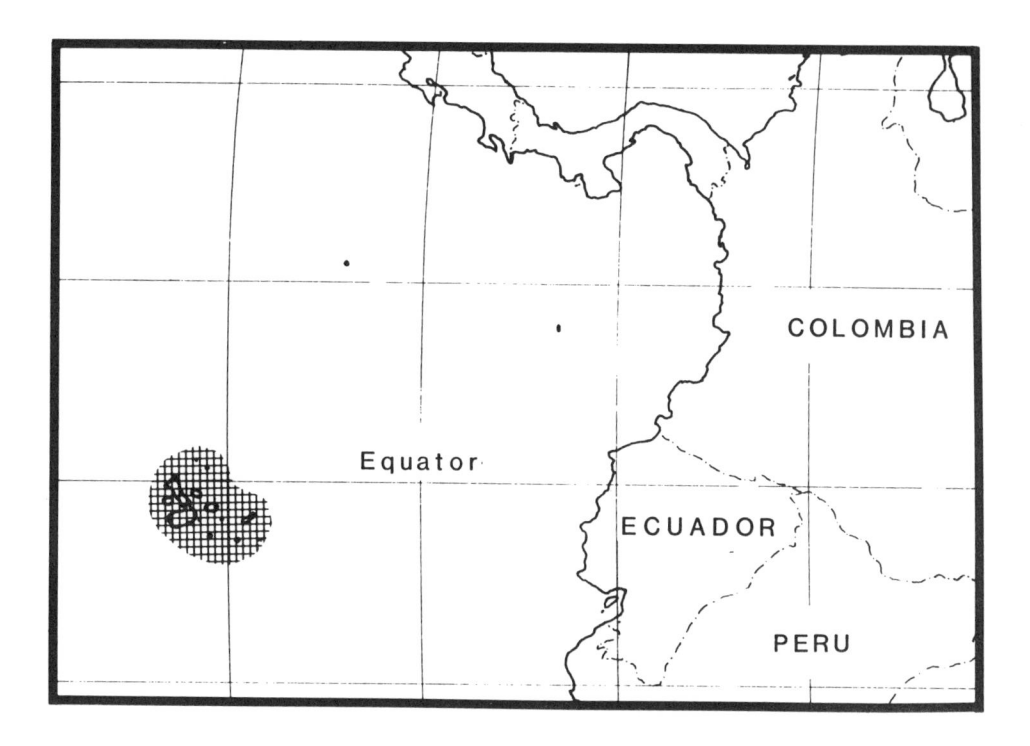

FIG. 57. Geographic distribution of Amblyrhynchina (*Amblyrhynchus* and *Conolophus*).

Distribution: Islands of the Galápagos Archipelago, Ecuador (Fig. 57).

Diagnosis: Members of this taxon can be distinguished from other iguanines by the following synapomorphies:

1. Nasal process of premaxilla covered dorsally between nasals (5-B).

2. Prefrontal contacts jugal, and lacrimal fails to contact palatine behind lacrimal foramen (7-B). This character occurs also in some *Ctenosaura* (*clarki*) and *Cyclura* (*carinata*, *cornuta*, and *ricordii*), in which it is interpreted as at least two separate instances of convergence with the condition seen in Amblyrhynchina.

3. Frontal wider than long (8-B). This character occurs also in *Cyclura cornuta* and *Iguana delicatissima*, which I interpret as two separate instances of convergence.

4. Lacrimal relatively small (17-B,-C).

5. Dorsal surface of vomerine process of palatine bears a high medial crest (21-B).

6. Labial process of coronoid relatively large (36-B,-C). This character occurs also in *Brachylophus*, in which it is interpreted as convergent.

7. Angular does not extend up lateral surface of mandible and is barely visible in lateral view (37-B). This character also occurs in *Sauromalus* and is either convergent or a synapomorphy of a more inclusive taxon.

8. Surangular not exposed, or only barely exposed on lingual surface of mandible between ventral processes of coronoid (40-B). This character occurs also in *Cyclura cychlura*, in which it is interpreted as convergent; it also occurs as a rare variant in several other iguanines.

9. Premaxillary teeth have large lateral cusps (45-B).

10. Anterior portion of pterygoid tooth patch absent (50-B). The entire pterygoid tooth patch is absent in most *Conolophus* and *Dipsosaurus*; however, when present, the pterygoid teeth of *Dipsosaurus* lie along the medial border of the pterygoid, those of *Conolophus* are located more laterally.

11. Second ceratobranchials relatively short, often less than two-thirds the length of the first ceratobranchials (52-53-A). This character also occurs in *Sauromalus* and is either convergent or a synapomorphy of a more inclusive taxon.

12. Caudal autotomy septa absent (60-B). This character also occurs in *Brachylophus* and in *Iguana delicatissima,* in which it is interpreted as two separate instances of convergence.

13. Dorsal head scales pointed and conical (83-B).

Fossil record: Steadman (1981) referred to *Conolophus* fossils of undetermined age from a lava tube on Isla Santa Cruz, Galápagos.

Comments: A close phylogenetic relationship between *Amblyrhynchus* and *Conolophus* is widely accepted (Heller, 1903; Eibl-Eibesfeldt, 1961; Avery and Tanner, 1971; Thornton, 1971; Etheridge in Paull et al., 1976) but supporting evidence other than geographic distribution has been scarce. Often the proposed close relationship between these taxa was merely asserted or based on unspecified similarities. Avery and Tanner (1971) did not distinguish between ancestral versus derived characters and used a highly artificial system for assessing similarity (see Introduction). The immunological studies of Higgins and Rand (1974, 1975; Higgins, 1977) compared the Galápagos iguanas only with *Iguana iguana* among iguanines. Wyles and Sarich (1983) performed more extensive immunological comparisons, including outgroups, but they prepared antisera to only four of the ten iguanine taxa used in their study. The morphological data presented in this study support the view that *Amblyrhynchus* and *Conolophus* are one another's closest living relatives.

The relationships of Amblyrhynchina to other Iguanini are uncertain. Although members of Amblyrhynchina share two derived characters with *Sauromalus*-reduced lateral exposure of the angular (37-B) and short second ceratobranchials (52-A)-I do not consider this convincing evidence for a close relationship between these taxa. At least one other character, highly cuspate marginal teeth (46-B,C), suggests a close relationship among *Sauromalus, Iguana,* and *Cyclura.* Convergences between *Amblyrhynchus* and *Sauromalus* are discussed in the comments on *Sauromalus*, above.

Amblyrhynchus Bell 1825

Type species (by monotypy): *Amblyrhynchus cristatus* Bell 1825.

Etymology: (Greek) *Amblys*, blunt, + *rhynchos*, snout.

Definition: The most recent common ancestor of the populations of Recent *Amblyrhynchus cristatus* and all of its descendants.

Distribution: Rocky coasts of islands in the Galápagos Archipelago, Ecuador (Fig. 57).

Diagnosis: Members of this taxon can be distinguished from other iguanines by the following synapomorphies:
 1. Posterolateral processes on ventral surface of premaxilla absent (1-B).
 2. Anterior surface of premaxillary rostral body nearly flat (3-B).
 3. Nasal process of premaxilla nearly vertical (4-B).
 4. Nasal capsule greatly inflated; nasal bones relatively large (6-B).
 5. Frontal develops deep, paired pockets on ventral surface (11-B).
 6. Dorsal orbital borders wedge-shaped (12-B).
 7. Maxilla flares outward below row of supralabial foramina (15-B).
 8. Lacrimal very small (17-C).
 9. Large ventral process of squamosal (18-A). Because a reduced ventral process is interpreted as a synapomorphy of Iguaninae, this is a character reversal. A similar character in *Iguana* is interpreted as convergent.
 10. Anterodorsal surface of septomaxilla bears pronounced longitudinal crest (20-B).
 11. Parasphenoid rostrum very short (27-B).
 12. Stapes relatively thick (31-B).
 13. Dorsal edge of dentary much higher than dorsal edge of surangular on either side of coronoid (32-B).
 14. Anterior inferior alveolar foramen located between coronoid and splenial; dentary does not contribute to its border (34-35-C).
 15. Angular process of prearticular remains relatively small throughout ontogeny (41-B).
 16. Crowns of posterior marginal teeth tricuspid (46-A). This is a reversal, since the presence of four or more cusps on the posterior marginal teeth is interpreted as a synapomorphy of Iguanini or possibly a more inclusive group. The presence of tricuspid posterior marginal teeth in adult *Ctenosaura bakeri* and *C. quinquecarinata* is interpreted as convergent.
 17. Secondary cusps of tricuspid marginal teeth relatively large, only slightly smaller than apical cusp (47-B).

18. Second ceratobranchials of hyoid apparatus separated from one another medially for most or all of their lengths (54-B). This character occurs also in *Sauromalus*, in which it is interpreted as convergent.

19. Scapular fenestrae small or absent (65-B). This character occurs also in *Sauromalus*, in which it is interpreted as convergent.

20. Posterior process of interclavicle does not extend beyond lateral corners of sternum (68-B). This character occurs also in *Sauromalus*, in which it is interpreted as convergent.

21. Lateral processes of interclavicle form angles of between 75° and 90° with the posterior process; interclavicle roughly T-shaped (69-B). This character also occurs in *Sauromalus*, in which it is interpreted as convergent.

22. Sternal fontanelle small or absent (70-B). This character also occurs in *Sauromalus*, in which it is interpreted as convergent.

23. Xiphisterna separated from one another medially (71-B). The xiphisterna of *Sauromalus* are also separated medially but to a much greater extent. I consider this similarity to be convergent.

24. Separable skull osteoderms develop over frontal, prefrontal, and nasal bones (74-B).

25. Colic wall without valves but with numerous irregular transverse folds (79-B).

26. Superciliary scales quadrangular and nonoverlapping (84-A). This character also occurs in *Sauromalus*, in which it is interpreted as convergent.

27. Gular fold weakly developed (87-B).

28. Digits of manus and pes partially webbed (93-B).

Other possible synapomorphies of *Amblyrhynchus* are a laterally compressed tail (Tracy and Christian, 1985) and a high rate of tooth replacement associated with wide alveolar margins of the maxilla, premaxilla, and dentary.

In addition, the following derived character occurs in some *Amblyrhynchus*:

Infraorbital foramen located entirely within palatine bone (24-B). This character occurs also in *Brachylophus* and in some *Ctenosaura* and *Sauromalus*.

Fossil record: None.

Comments: Monophyly of *Amblyrhynchus* is the best-supported phylogenetic hypothesis within Iguaninae. *Sauromalus* has almost as many characters that are interpreted as synapomorphies, but it has more that require convergence elsewhere within iguanines and, in this sense, are ambiguous. In terms of a simple tally of derived characters used in this study, *Amblyrhynchus* is the most highly modified iguanine relative to the most recent common ancestor of them all. *Amblyrhynchus* not only possess numerous synapomorphies supporting its own monophyly, but also possesses those of Amblyrhynchina and Iguanini. This high degree of morphological modification is not surprising, given the unique natural history of these animals; *Amblyrhynchus* are the only extant lizards that gain a major part of their sustenance from the sea (Darwin, 1835; Heller, 1903; Carpenter, 1966; Dawson et al., 1977; Dee Boersma, 1983). Many of the unique morphological features of *Amblyrhynchus* discussed in this paper are probably related to this unique mode of existence. For example, the modifications of the teeth and colon may be related to the unique diet of these lizards, which consists largely of marine algae (Darwin, 1835; Carpenter, 1966; Dee Boersma, 1983). Derived characters obviously associated with aquatic locomotion include the webbed digits and the strongly compressed tail. The thickened stapes may be related to differences between the sound-transmitting properties of water and air. The inflated nasal capsule and the deep pockets that develop on the ventral surface of the frontal house enlarged nasal salt glands, which allow marine iguanas to excrete excess salt accumulated from ingesting food with a salt concentration similar to that of seawater (Schmidt-Nielsen and Fange, 1958). Convergences between *Amblyrhynchus* and *Sauromalus* are discussed in the comments on the latter taxon, above.

Conolophus Fitzinger 1843

Type species (by original designation): *Amblyrhynchus demarlii* Duméril and Bibron 1837 = *Amblyrhynchus subcristatus* Gray 1831b.

Etymology: (Greek) *Konos*, cone, + *lophos*, crest, presumably referring to the conical scales of the dorsal crest.

Definition: The most recent common ancestor of *Conolophus pallidus* and *C. subcristatus* and all of its descendants.

Distribution: Islands of the Galápagos Archipelago, Ecuador (Fig. 57).

Diagnosis: Members of this taxon can be distinguished from other iguanines by the following synapomorphies:
  1. Lateral crests of premaxillary incisive process large and pierced or notched by foramina for maxillary arteries (2-B).
  2. Crista cranii of frontal projects anteriorly forming a step rather than a smooth curve where it meets medial edge of prefrontal at dorsal margin of orbitonasal fenestra (10-B).
  3. Supratemporals relatively small, extend one-half or less the distance across posterior temporal fossae (14-B).
  4. Ectopterygoid contacts palatine near posteromedial corner of suborbital fenestra (26-B). This character occurs also in about half of the *Iguana delicatissima* examined, in which it is interpreted as convergent.
  5. Labial process of coronoid very large, extends more than two-thirds the way down lateral surface of mandible in large specimens (36-C).
  6. Pterygoid teeth usually absent (51-B). This character also occurs in *Dipsosaurus*; however, when present, the pterygoid teeth of *Dipsosaurus* lie along the medial border of the pterygoid while those of *Conolophus* are situated more laterally.

7. Foramina in ventral surface of second sacral pleurapophyses usually absent; their place taken by open grooves (58-B). This character exists as a polymorphism in both species of *Conolophus*; that is, it does not characterize all specimens.

8. Subclavian arteries not covered ventrally by *M. rectus capitis anterior* (76-B). This character needs to be checked in additional specimens.

Fossil record: Steadman (1981) reported *Conolophus* fossils of undetermined age from a lava tube on Isla Santa Cruz, Galápagos.

Etheridge (1964b) reported a fragmentary braincase and a body vertebra from Late Pleistocene cave deposits on the West Indian island of Barbuda and estimated that both were from animals about 400 mm snout-vent length. He stated that the body vertebra, with its robust neural spine and well developed zygosphenes and zygantra, is similar to those of large iguanines. Etheridge compared the braincase with those of various large iguanines noting, as pointed out by Boulenger (1890), that the parabasisphenoid "is much wider than long and slightly to moderately constricted behind the [basi]pterygoid processes in *Iguana* and *Cyclura*, about as wide as long and strongly constricted in *Amblyrhynchus* and *Conolophus*, and much longer than wide and strongly constricted in *Ctenosaura*" (Etheridge, 1964b:68). He also gave the following length-to-width ratios for the parabasisphenoid (length measured from posterior border to apex of indentation between basipterygoid processes, width measured at narrowest point posterior to basipterygoid processes): *Iguana* .40-.65, *Cyclura* .64-.72, *Amblyrhynchus* .79-.91, *Conolophus* .86-1.10, *Ctenosaura* 1.45-1.96. Because the ratio of the fossil is 1.00, Etheridge concluded that it most closely resembles *Conolophus*.

Based on its large size and the presence of zygosphenes and zygantra, the vertebra is reasonably interpreted as belonging to an iguanine. Based on size, both vertebra and braincase might tentatively be referred to Iguanini. The similar proportions of the parabasisphenoid in the fossil and *Conolophus*, however, provide no evidence that the two are closely related. The wide parabasisphenoids of *Iguana* and *Cyclura,* and the long one of *Ctenosaura,* are derived conditions, while *Amblyrhynchus*, *Conolophus*, and the fossil retain primitive proportions of this element. Furthermore, the proportions of the parabasisphenoid in the fossil fall within the range of variation not only of *Conolophus* but also of *Cyclura*. Etheridge's (1964b) range of .64-.72 for the length/width of the parabasisphenoid in *Cyclura* is based on *C. cornuta*, *C. figginsi* (=*cychlura*), *C. ricordii*, and *C. macleayi* (=*nubila*), but the range is actually much greater when other *Cyclura* are included. Pregill (1981) reported a ratio of .50 for a Puerto Rican fossil that he referred to *C. pinguis*, and I have obtained a range of .52 (*C. pinguis*) to 1.10 (*C. carinata*). Although neither *Conolophus* nor *Cyclura* occurs on Barbuda today, *Cyclura* occurs in the West Indies, while *Conolophus* is restricted to the Galápagos Islands. Nevertheless, current knowledge does not permit me to refer the Barbuda fossil to either of these taxa.

Comments: Although not as obviously modified from the ancestral Amblyrhynchina as its sister taxon, *Amblyrhynchus*, *Conolophus* has eight derived characters not seen in

Amblyrhynchus. These characters indicate that *Conolophus* is monophyletic and thus, contrary to one commonly entertained hypothesis about the relationships of the Galápagos iguanas (Thornton, 1971; Higgins, 1978), cannot be considered ancestral to *Amblyrhynchus*.

Iguanina Bell 1825

Type genus: Iguana Laurenti 1768.

Etymology: Modification of *Iguana*, the name of its type genus.

Definition: The most recent common ancestor of *Cyclura* and *Iguana* and all of its descendants.

Distribution: Lowlands of the American mainland from Sinaloa and Veracruz, México, southward through Central America and northern South America to southern Brazil and Paraguay as well as various Caribbean islands, including both the Greater and Lesser Antilles.

Diagnosis: Members of this taxon can be distinguished from other iguanines by the following synapomorphies:
1. Squamosal abuts against dorsal end of tympanic crest of quadrate (19-B).
2. Cristae ventrolaterales of parabasisphenoid only narrowly constricted behind basipterygoid processes (28-B,-C). This character does not occur in *Cyclura carinata*.
3. Surangular exposed laterally well anterior to apex of coronoid and often anterior to last dentary tooth (39-B). This character also occurs in some *Ctenosaura* and may be a synapomorphy of a more inclusive group.
4. Crowns of posterior marginal teeth with five or more cusps (46-C,-D). This character occurs also in *Sauromalus* and may be a synapomorphy of a more inclusive group.

Another possible synapomorphy of Iguanina is the development of a dewlap. Although the dewlap of *Cyclura* is relatively small compared with that of *Iguana*, it is larger than that of other iguanines except *Brachylophus* and some *Ctenosaura*.

Fossil record: The oldest known fossil referable to Iguanina is an almost complete skull from the Pliocene of southern California (Norell, 1983). This and other fossil Iguanina are discussed further in the sections on the fossil records of *Iguana* and *Cyclura*, below.

Comments: The name Iguanina is first used in this work; Bell (1825) is credited with authorship under the principle of coordination (Article 36, third edition of the *International Code of Zoological Nomenclature*).

Although there are fewer characters supporting a sister-group relationship between *Cyclura* and *Iguana* than there are for some of the other relationships proposed in this paper, the monophyly of Iguanina is reasonably well supported. This sister group of Iguanina is not obvious from the results of the present study, but the best candidates are *Ctenosaura* and *Sauromalus* (or perhaps a group composed of both these taxa). Only *Sauromalus* shares a derived character with Iguanina that is not variable within either of these taxa, increased cuspation of the posterior marginal teeth (46-B,-C). The distribution of other derived characters among taxa within Iguanini requires either that one or more of the basic taxa are not monophyletic or that some kind of homoplasy is involved.

Iguana Laurenti 1768

Type species (by tautonomy): *Lacerta iguana* Linnaeus 1758.

Etymology: (Spanish) *Iguana*, a modification of the name given to these animals by West Indian natives.

Definition: The most recent common ancestor of *Iguana delicatissima* and *I. iguana* and all of its descendants.

Distribution: Lowlands of the American mainland from Sinaloa and Veracruz, México, southward through Central America and northern South America to southern Brazil and Paraguay; in the Caribbean northward through the Lesser Antilles to the Virgin Islands (Fig. 58).

Diagnosis: Members of this taxon can be distinguished from other iguanines by the following synapomorphies:
 1. Large ventral process of squamosal (18-A) abuts against dorsal edge of tympanic crest of quadrate. Because the reduction of the ventral process of the squamosal is an iguanine synapomorphy, its reelaboration in *Iguana* is a character reversal.
 2. Cristae ventrolaterales of parabasisphenoid barely constricted behind basipterygoid processes (28-C). This character also occurs in some *Cyclura* and, although I have interpreted this as convergence, a wide parabasisphenoid may be a synapomorphy of a more inclusive group.
 3. Crowns of posterior marginal teeth serrate, with numerous small accessory cusps (46-D).
 4. Entire pterygoid tooth patch doubles ontogenetically (49-C).
 5. Posterolateral processes of pleurapophyses of second sacral vertebra absent (57-B). This character occurs also in *Ctenosaura* and in most *Cyclura,* and may be a synapomorphy of a more inclusive group.

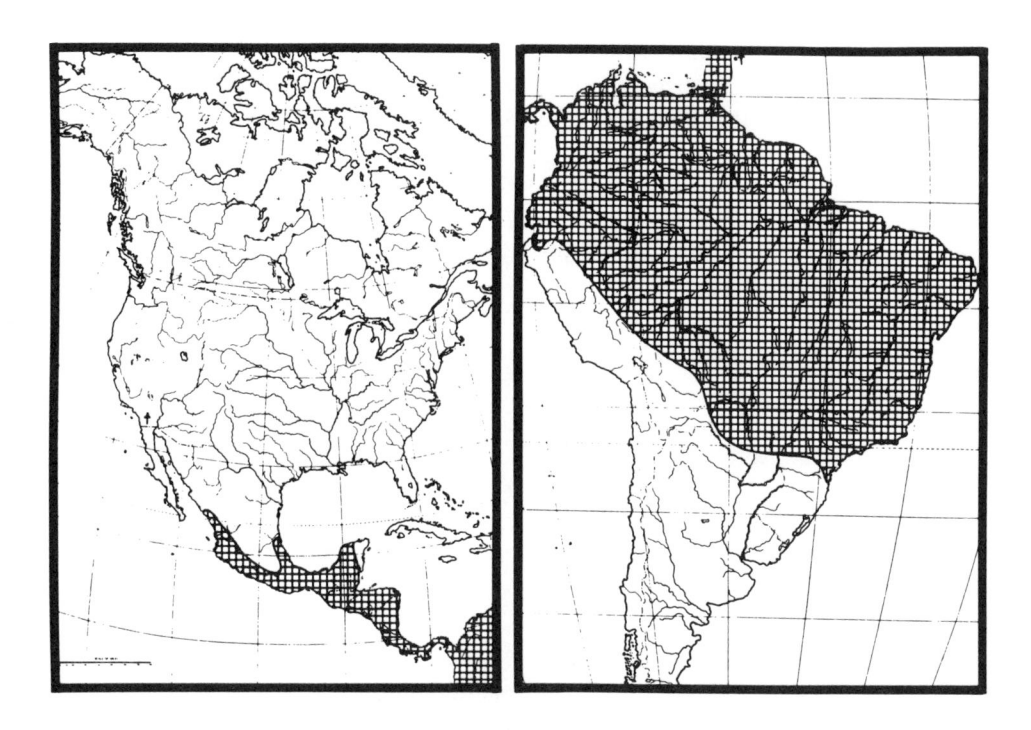

FIG. 58. Geographic distribution of *Iguana* (from Etheridge, 1982).

6. Thin midsagittal processes on anterodorsal surfaces of caudal centra relatively small and confined to anterior fifth of caudal sequence (62-B). This character occurs also in *Brachylophus,* in which it is interpreted as convergent.

7. Coeliac artery arises posterior to mesenteric arteries, between mesenteric arteries, or continuous with mesenteric arteries rather than anterior to them (78-B). This character also occurs in some *Sauromalus,* in which it is interpreted as convergent. It needs to be checked in larger samples.

8. Snout covered by large, platelike scales (82-B). This character also occurs in some *Cyclura* and may be a synapomorphy of a more inclusive group.

9. Large dewlap present (88-B). This character also occurs in *Brachylophus* and in *Ctenosaura bakeri* and *C. palearis,* in which it is interpreted as two separate instances of convergence.

10. Gular crest of enlarged scales present (88-B).

In addition, the following derived characters occur in only some *Iguana*:

1. Frontal wider than long (8-B). This character occurs also in Amblyrhynchina and in *Cyclura cornuta,* in which it is interpreted as two separate instances of convergence. Within *Iguana* it occurs only in *I. delicatissima* and appears to be a synapomorphy of that taxon.

2. Second ceratobranchials of hyoid apparatus much longer than first ceratobranchials (52-53-C). This character also occurs in *Brachylophus,* in which it is interpreted as convergent. Within *Iguana* it occurs only in *I. iguana* and appears to be a synapomorphy of that taxon.

3. Caudal autotomy septa absent (60-B). This character occurs also in Amblyrhynchina and in *Brachylophus,* in which it is interpreted as two separate instances of convergence. Within *Iguana* it occurs only in *I. delicatissima* and appears to be a synapomorphy of that taxon.

Fossil record: Fossils referred to *Iguana* have been reported from Antigua (Wing et al., 1968), Barbados (Swinton, 1937; Ray, 1964), Martinique (Hoffstetter, 1946), and Montserrat (Steadman et al., 1984) in the West Indies; Ecuador (Hoffstetter, 1970); and southern California (Norell, 1983). The oldest of these are the Pliocene specimens from southern California. However, since Norell (1983) considers these to be outside of the clade consisting of *I. iguana* and *I. delicatissima*, they are not *Iguana* according to my definition of this taxon, although they are its closest relatives. All other fossils referred to *Iguana* are either Upper Pleistocene or Holocene in age.

Comments: I consider the monophyly of *Iguana* to be reasonably well supported. Nevertheless, three of the derived characters employed in this study occur in some *Iguana* as well as in either *Brachylophus* (character 52-53-C), Amblyrhynchina (8-B), or both of these taxa (60-B). The reason that I have interpreted these characters as convergent is acceptance of the monophyly not only of *Iguana* but also of Iguanina and Iguanini, based on other characters.

Within *Iguana*, the two currently recognized species both appear to be monophyletic; therefore, neither can be considered to be ancestral to the other. Monophyly of *I. iguana* is supported by the extreme width of the parabasisphenoid, the enlarged subtympanic scale (Dunn, 1934; Lazell, 1973), and the elongated second ceratobranchials. Monophyly of *I. delicatissima* is supported by the short frontal bone, absence of autotomy septa in the caudal vertebrae, enlarged bony external nares, and possibly the failure of the septomaxillae to reach the roof of the nasal capsule. The possibility that *Iguana* is a subgroup of of *Cyclura* is discussed in the comments on the latter taxon, below.

Cyclura Harlan 1824

Type species (subsequent designation by Fitzinger 1843): *Cyclura carinata* Harlan 1824.

Etymology: (Greek) *Kyklos*, circle, + *oura*, tail, referring to the verticils of enlarged, spinous scales on the tails of most species.

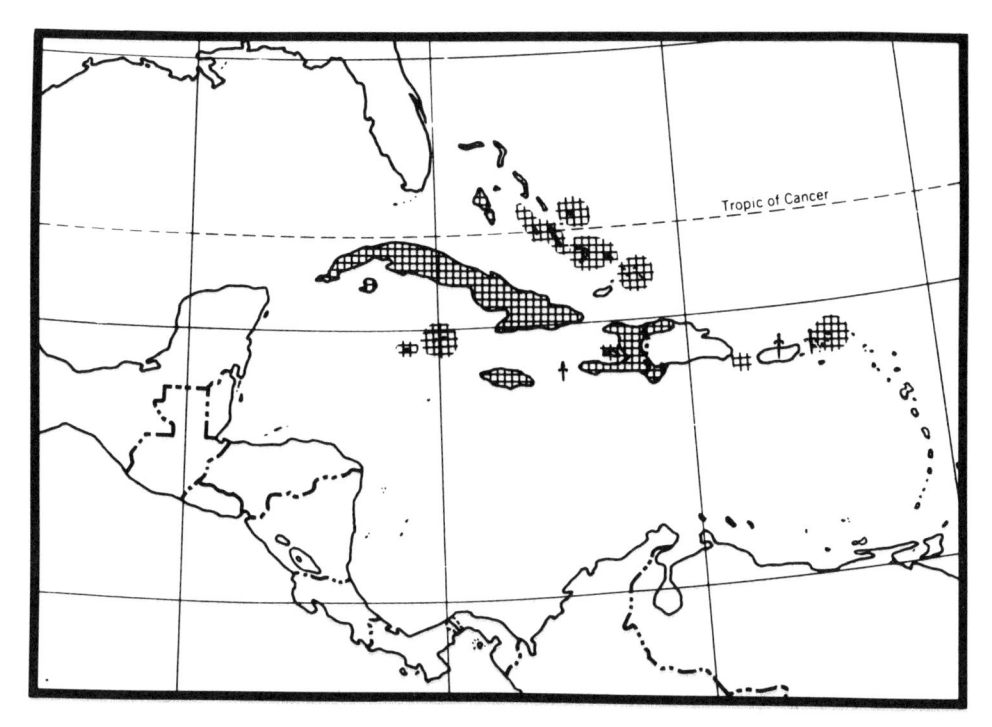

Phylogenetic Systematics of Iguanine Lizards

FIG. 59. Geographic distribution of *Cyclura* (from Schwartz and Carey, 1977).

Definition: The most recent common ancestor of Recent *Cyclura* (*carinata, collei, cornuta, cychlura, nubila, pinguis, ricordii,* and *rileyi*) and all of its descendants.

Distribution: The Bahama Islands; Cayman Islands; Mona and Anegada islands; and Cuba, Hispaniola, and Jamaica, and their nearby islets (Fig. 59). *Cyclura* is nearly extinct on Jamaica (Woodley, 1980) and has become extinct on Navassa Island in historical times (Thomas, 1966).

Diagnosis: Members of this taxon can be distinguished from other iguanines by the following synapomorphies:
 1. Modal number of premaxillary teeth greater than seven (43-44-C). This character applies only to populations; its presence or absence cannot always be inferred from the condition in a single organism.
 2. Presence of toe combs formed by enlargement of anterior keels of subdigital scales and fusion of their bases (92-B). *Ctenosaura defensor* also possesses enlarged and fused subdigital keels, which are interpreted as convergent; however, in this taxon, they occur only under the first phalanx of digit III. In *Cyclura* the enlarged and fused subdigital keels occur under the first phalanx of digit II and the first and second phalanges of digit III.
 In addition, the following derived characters occur only in some *Cyclura*:

1. Prefrontal contacts jugal and lacrimal fails to contact palatine behind lacrimal foramen (7-B). This character occurs also in Amblyrhynchina and in some *Ctenosaura*; within *Cyclura*, it occurs only in *C. carinata*, *C. cornuta*, and *C. ricordii*.

2. Frontal wider than long (8-B). This character occurs also in Amblyrhynchina and *Iguana delicatissima*, in which it is interpreted as convergent. Within *Cyclura* it occurs only in *C. cornuta* and appears to be a synapomorphy of this taxon.

3. Parietal foramen located variably or invariably within frontal bone (13-B,-C). This character occurs also in *Sauromalus*, *Dipsosaurus*, and some *Ctenosaura*. Within *Cyclura*, invariable location of the parietal foramen within the frontal is characteristic only of *C. carinata*.

4. Cristae ventrolaterales of parabasisphenoid barely constricted behind basipterygoid processes (28-C). This character occurs also in *Iguana*. Within *Cyclura*, it varies considerably among taxa; in some (e.g., *C. pinguis*) the ventral surface of the parabasisphenoid is as wide or wider than that of *I. delicatissima*, while in others (e.g., *C. carinata*) it is relatively narrow, though still wider than in most iguanines other than *Iguana* and other *Cyclura* (see section on fossil record of *Conolophus*, above).

5. Surangular not exposed or only barely exposed below coronoid on lingual surface of jaw (40-B). This character occurs also in Amblyrhynchina, in which it is interpreted as convergent. Within *Cyclura*, it characterizes only *C. cychlura*, although it occurs at a moderate frequency in *C. nubila*.

6. Posterior portion of pterygoid tooth patch doubles ontogenetically (49-B). Within *Cyclura*, this character occurs in *C. cornuta*, *C. nubila*, *C. pinguis*, and *C. ricordii*. This character, or a further modification of it, occurs only in *Iguana* and some *Ctenosaura* and *Cyclura*. Because its expression seems to depend on size, posterior doubling of the pterygoid tooth patch may be a synapomorphy of a more inclusive group, in which case failure to double would be a synapomorphy within *Cyclura*.

7. Posterolateral processes of pleurapophyses of second sacral vertebra absent (57-B). This character occurs also in *Ctenosaura* and *Iguana* and may be a synapomorphy of a more inclusive group. Within *Cyclura*, I have found the processes only in *C. pinguis*.

8. Snout covered by large, platelike scales (82-B). This character occurs also in *Iguana*. Within *Cyclura*, it occurs in all taxa except *C. carinata* and *C. ricordii*. Considerable variation in the size of these scales exists even among those *Cyclura* possessing enlarged snout scales (figures in Schwartz and Carey, 1977).

9. Tail bears verticils of enlarged, spinous scales (94-B). This character occurs also in *Ctenosaura*. The degree of caudal spinosity exhibits considerable variation within *Cyclura* (figures in Barbour and Noble, 1916).

Fossil record: Fossil *Cyclura* have been reported from the Upper Pleistocene and Holocene of Puerto Rico (Barbour, 1919; Pregill, 1981), St. Thomas in the Virgin Islands (Miller, 1918), and New Providence Island in the Bahamas (Etheridge, 1965c; Pregill, 1982). The specimens from Puerto Rico have been referred to the extant species *C. pinguis* (Pregill, 1981). A braincase and a body vertebra from the Late Pleistocene of

Barbuda may also be remains of *Cyclura* (Pregill, 1981; see section on the fossil record of *Conolophus*, above).

Comments: *Cyclura* is often assumed to be closely related to *Ctenosaura* (Barbour and Noble, 1916; Bailey, 1928; Schwartz and Carey, 1977), which it resembles in general body form, terrestrial habits, and the verticils of enlarged, spinous caudal scales. These similarities were noticed at least as early as Harlan (1824), who erected *Cyclura* for species that are now placed in both *Cyclura* (*carinata*) and *Ctenosaura* (*teres* = *acanthura*). Despite the resemblance between *Ctenosaura* and *Cyclura*, *Cyclura* probably shared a more recent common ancestor with *Iguana* than it did with *Ctenosaura*. The similarities between *Cyclura* and *Ctenosaura* in general body form and terrestriality probably represent primitive features retained from the common ancestor of all three taxa; and since not all *Cyclura* possess the verticils of enlarged, spinous caudal scales, some form of homoplasy in tail morphology is required no matter which relationships are accepted. Furthermore, *Cyclura* and *Iguana* share at least three derived characters not seen in *Ctenosaura*: abutment of the squamosal against the dorsal end of the tympanic crest of the quadrate (19-B); a wide parabasisphenoid (28-B,-C); and highly cuspate posterior marginal teeth (46-C,-D). Although the last character occurs also in *Ctenosaura defensor*, my analysis of relationships within *Ctenosaura* indicates that this is convergent.

In addition to the characters suggesting a close phylogenetic relationship between *Iguana* and *Cyclura*, there are other characters suggesting that *Iguana* is actually a subgroup of *Cyclura*, as defined here. In other words, there are characters suggesting that the most recent common ancestor of all *Cyclura* was also an ancestor of *Iguana*. *Iguana* shares derived features of the cephalic scutellation, such as the enlarged snout scales and the row of enlarged sublabials, as well as a derived widening of the parabasisphenoid with some, but not all, species of *Cyclura*. There is a particularly close resemblance between *Cyclura cychlura* and *Iguana delicatissima* in these features. Nevertheless, the toe combs, the verticils of enlarged, spinous caudal scales, and the high number of premaxillary teeth are derived features seen in *Cyclura* but not in *Iguana*. The morphology of the posterior marginal teeth also varies within *Cyclura*, with some approaching the highly cuspate morphology seen in *Iguana* much more closely than others. However, the high degree of variation in this character, at least some of which is ontogenetic, along with small samples and ambiguities caused by wear, prevent me from making any definite statement about the relationships suggested by this character.

In any case, the precise relationships between *Iguana* and *Cyclura* are unclear, because the distributions of derived characters among taxa contradict one another. I provisionally accept the monophyly of *Cyclura*, but consider the issue to be in need of further study. If the most recent common ancestor of all *Cyclura* was also an ancestor of *Iguana*, then, according to the phylogenetic definitions of taxa adopted here, *Iguana* is a subgroup of *Cyclura* rather than a separate taxon, and Iguanina is a synonym of *Cyclura*.

Appendix I
Specimens Examined

All specimens listed below are partial or complete skeletons unless otherwise indicated as alcoholic specimen (A) or radiograph (R). Institutional abbreviations are as follows:

AMNH, American Museum of Natural History

ASFS, Collection of Albert Schwartz (Miami-Dade Community College North, Miami, Fla.)

CAS, California Academy of Sciences

JMS, Collection of Jay M. Savage (University of Miami, Coral Gables, Fla.)

KdQ, Collection of the author (University of California, Berkeley)

KU, University of Kansas Museum of Natural History

LACM, Los Angeles County Museum of Natural History

LSUMZ, Louisiana State University, Museum of Zoology

MCZ, Museum of Comparative Zoology, Harvard University

MVZ, Museum of Vertebrate Zoology, University of California, Berkeley

RE, Collection of Richard Etheridge (San Diego State University, San Diego, Calif.)

SDNHM, San Diego Natural History Museum

UCMP, Museum of Paleontology, University of California, Berkeley

UF, Florida State Museum, University of Florida

USNM, United States National Museum of Natural History.

IGUANINES

Amblyrhynchus cristatus: JMS 126-7, 181, 222; LACM 127324; RE 338, 386, 1041, 1091, 1095, 1196, 1387, 1396, 1508, 2239; SDNHM 45156-7, 47000, 55600.

Brachylophus fasciatus: AMNH 17701; CAS 54664 (A); RE 1019, 1770, 1866, 1888, 2089, plus two radiographs of specimens whose institutions of deposition are unknown; MCZ 5222, 5800, 15008-9; SDNHM 55289, 55601, 55603, 60429, 62341.
B. vitiensis: MCZ 158238, 160253-5.

Conolophus pallidus: JMS 61, 213-8; MCZ 79772; RE 439-40, 1382, 1446-7.
C. subcristatus: AMNH 50798, 71304, 110167-8, 114493; CAS 12058 (A); MCZ 2027; MVZ 77314; RE 327; SDNHM 33682, 47007, 47140; USNM 89992, 165756.
C. sp.: AMNH 14494.

Ctenosaura acanthura: AMNH 46483; MCZ 2176, 5013-21, 11350; SDNHM 47004, 59542-3; USNM 220217-8.
C. bakeri: ; LSUMZ 22275, 22293, 22367-71, 22399 (all A,R); UF 28530-33 (A,R; 28530 also skeletonized); USNM 25324 (skull drawing from Ray and Williams, unpubl.).
C. clarki: JMS 1544; MCZ 22454; MVZ 76690 (A,R), 76694 (A,R), 79256, 79293, 164865-66; RE 57, 184; USNM 21450.
C. defensor: KU 70261-2, 75528 (all A,R); MCZ 7095 (skull, A, and R); UF 41534 (A,R).
C. hemilopha: JMS 287-9, 291, X366 (R), X631-2 (R), X634-5 (R); RE 325, 491, 497-8, 502, 1087, 1341, 1386, 1887, 1964; SDNHM 48480, 48976, 55290, 57114.
C. palearis: CAS 69297, 69299, 69307, 69310 (all A,R); MCZ 22390, 22399; MVZ 162073-5, 162305 (all A,R).
C. pectinata: JMS 238, 242, 250, 269, 692, 696, 704, 1252; RE 56, 419-21, 490, 641; SDNHM 55291.
C. quinquecarinata: AMNH 77640; CAS 73554-62 (A,R); MCZ 24903; MVZ 79294, 128903 (A,R).
C. similis: AMNH 38949; JMS 178; MCZ 5011, 5457, 5799, 9566, 10312, 21742, 22662, 25993, 26968, 27207, 36830, 139421; RE 469, 2003, 2233, 2238.

Cyclura carinata: CAS 54647 (A); JMS 98; MCZ 59255, 139424; RE 1969; USNM 88819.
C. collei: CAS 74731 (A); MCZ 9397 (R).
C. cornuta: AMNH 57878, 114487-8; JMS 221; MCZ 9974 (R); RE 383, 1226, 1837, 1841-2, 1858, 1962, 1981-2, 1991.
C. cychlura: AMNH 74440, 76875, 76877-8; KdQ 47-8; MCZ 6915; RE 2073; USNM 64650.
C. nubila: JMS 180, 182, 273; MCZ 6915; RE 228, 337, 610; SDNHM 42957, 42960.
C. pinguis: ASFS V21995.
C. ricordii: JMS 272, 367; RE 435.
C. rileyi: MCZ 38165-69 (A); RE uncatalogued (A); UF 40744 (A), 57741.

Dipsosaurus dorsalis: JMS X320, X324, X331, X334, X336, X338, X607, X612, X618 (all R); RE 33, 355-9, 484, 661, 667, 1497, 1572-7, 1848, 1868, 1980; SDNHM 47006, 57107-9, 59538-9, 60424.

Iguana delicatissima: KdQ 21; MCZ 6097, 10975, 16157, 60823, 75388, 83228.
I. iguana: JMS 244-5, 268, 713, 1028, 1545, 1553; RE 89, 158, 424, 452-4, 468, 489, 1006, 1850, 1886, 2232; SDNHM 47001, 47008, 47010, 59466, 59540-1.

Sauromalus ater: JMS 39; KdQ 68-9; MCZ 31521; RE 1504; SDNHM 6865.
S. australis: KdQ 71, 72.

S. hispidus: JMS 172-4, 219, 239, 401, 404, 436, 915, 983; LACM 127279; RE 317, 514-5, 736, 803-5, 1042, 1384, 1927, 1974; SDNHM 6873, 47028-30, 57103-4, 59471.
S. obesus: RE 244, 354, 380, 408-11, 426, 461-7, 1578-9, 1852, 1864; SDNHM 48483, 59534.
S. slevini: MCZ 85553; RE 1340, 1367.
S. varius: JMS 175-6, 246-8; RE 308, 323, 451, 512-3, 539, 1043, 1404, 1928, 2084; SDNHM 47024-6, 59542.

BASILISCINES

Basiliscus basiliscus: JMS 347, 362, 1449, 1567, 1577, 1583-4; RE 555.
B. plumifrons: RE 427, 2014; SDNHM 57098-100, 59467, 60430-1.
B. vittatus: RE 49, 637, 1601, 1729, 1757, 1759, 2015; SDNHM 60432.

Corytophanes cristatus: JMS 1701; KdQ 55; SDNHM 62345.
C. hernandesii: RE 1176, 1800.

Laemanctus longipes: MVZ 137673; UCMP 129880.
L. serratus: AMNH 44982; RE 619.

CROTAPHYTINES

Crotaphytus collaris: RE 85, 370-1, 404-7, 683, 1213-4, 1570, 1797, 1823, 1836, 1857; SDNHM 60433-5.
C. insularis: SDNHM 47002.

Gambelia wislizenii: RE 425, 550, 810, 1029, 1172, 1571.

MORUNASAURS

Enyalioides heterolepis: AMNH 18232, 18278-9 (all R); MCZ 8063, 24959, 39977 (all R).
E. laticeps: AMNH 37561 (R); MCZ 37282, 37284, 37286, 50238 (all R); RE 76, plus three radiographs of specimens whose institutions of deposition are unknown; SDNHM 47003.
E. microlepis: AMNH 37562, 60608-9 (all R).
E. oshaughnessyi: AMNH 28869-70, 28874-6, 28894 (all R); MCZ 29297 (R); RE 1957.
E. palpebralis: AMNH 56401, 57159-61 (all R); MCZ 84035 (R).
E. praestabilis: ANMH 37554-5 (R); USNM 7796-8 (R), 222583.

Hoplocercus spinosus: AMNH 90658, 93807; CAS 93081-94, 93804-5, 101443-5 (all R); RE 1263, 1502.

Morunasaurus annularis: AMNH 57178, 57180-2, 57199 (all R); MCZ 146375; RE 1956; USNM 616-7, 3782 (all R).
M. groi: CAS 98001 (R).

OPLURINES

Chalarodon madagascariensis: RE 455, 457, 547.

Oplurus cuvieri: JMS 330; MVZ 117597 (A,R), 128904 (A,R); RE 558, 620, 1835.
O. quadrimaculatus: AMNH 71452; RE 658.

In addition, I have examined various alcoholic specimens in the collections of Richard Etheridge, at San Diego State University; and the Museum of Vertebrate Zoology, University of California, Berkeley.

Appendix II

Polarity Determination Under Uncertain Outgroup Relationships

I used a modified version of the outgroup method described by Maddison et al. (1984) and M. J. Donoghue (pers. comm., 1982) to assess character polarities. This method assesses the condition of the outgroup node (branch point linking the ingroup with its sister group on a phylogenetic tree or a cladogram) in order to minimize character-state changes at all hierarchical levels. Briefly, the cladogram for the ingroup and various outgroups is rerooted at the outgroup node, and the conditions of the various subterminal nodes on the rerooted cladogram are assessed using an optimization procedure similar to that of Farris (1970). First, the terminal nodes (ends of branches) are labeled according to the condition found in the outgroup occupying that position. Second, the subterminal nodes are assigned character states according to the following rules: (1) If both nodes above the node in question have the same state, assign that state to the node in question. (2) If the two nodes above the node in question have different states, the assignment of the node in question is equivocal (?). (3) If one node above the node in question is equivocal and the other is not, assign the node in question the state of the unequivocal node. The state assigned to the outgroup node (basal node of the rerooted cladogram) is taken as plesiomorphic.

Because the relationships among iguanines and the outgroups used in this study are poorly understood, I was forced to consider all possible cladograms for four unspecified outgroups and an ingroup, of which there are nine (Fig. 60). After these cladograms are rerooted at the outgroup node (Fig. 61), it can be seen that not all of them need to be considered further, since many will yield identical assessments of the condition at the outgroup node. Complete equivalence is seen between some of the rerooted cladograms: A = E = G, and C = F. By swiveling branches about nodes, which does not alter the relationships implied by the diagrams, rerooted cladograms A, B, and D are found to be equivalent. Finally, given only the distribution of character states in the ingroup and these four outgroups, the state assigned to the outgroup node in rerooted cladograms H and I must be identical to that of the basal node in the clade formed by the four outgroups. Therefore, for the purposes of this analysis, rerooted cladograms H and I can be considered to be equivalent to A and C, respectively. Only two topologies need to be considered further, A and C (Fig. 61).

For any given character, the conditions of the outgroups can be placed on the terminal branches of the two rerooted cladograms (Fig. 61A, C) in all possible combinations, and the condition of the outgroup node (i.e., the character's polarity) can be assessed. For cases in which all four outgroups suggest a single interpretation, that interpretation is accepted. For cases in which more than one of the states found in the ingroup also occur in one or more outgroups, the polarity of the character is ambiguous. In such cases, I have

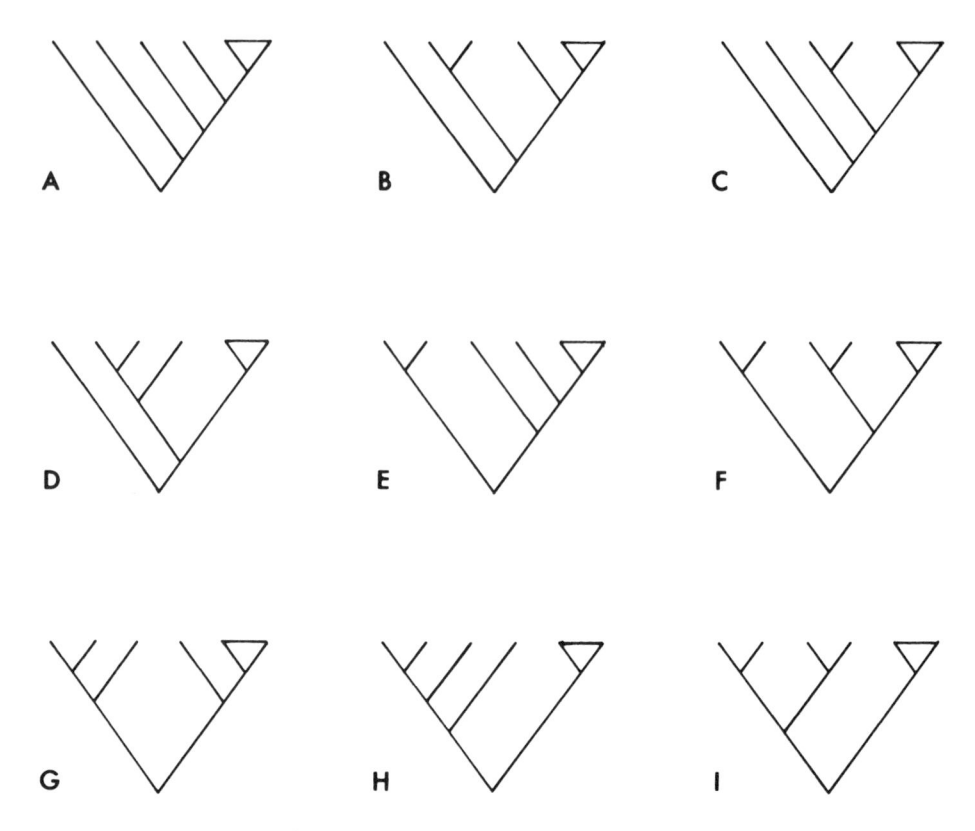

FIG. 60. All nine possible fully resolved cladogram topologies for four unspecified outgroups and an ingroup (inverted triangle).

made a compromise between maximizing the total number of characters on the one hand and using only those characters whose polarities are completely unambiguous on the other.

Table 11 shows polarity inferences for all possible arrangements of four outgroups on the two rerooted cladograms (Fig. 61A,C) for seven cases of character-state distribution. This exhausts the possible character-state distributions for two state characters, since it is the occurrence of a given state rather than its alphabetic designation that is important (e.g., A/A/A/B = B/B/B/A). The following is a case-by-case discussion of possible polarity inferences under different relationships of the four outgroups to the ingroup.

Case I (A/A/A/A,B): For the case in which three outgroups have one condition and the other has both alternative conditions, all arrangements except one require that the common state be considered plesiomorphic. The lone exception is when the variable outgroup attaches directly to the basal node of the rerooted cladogram (Fig. 62A). If resolution of relationships within this outgroup requires that state B be considered plesiomorphic for this group, the polarity will be equivocal.

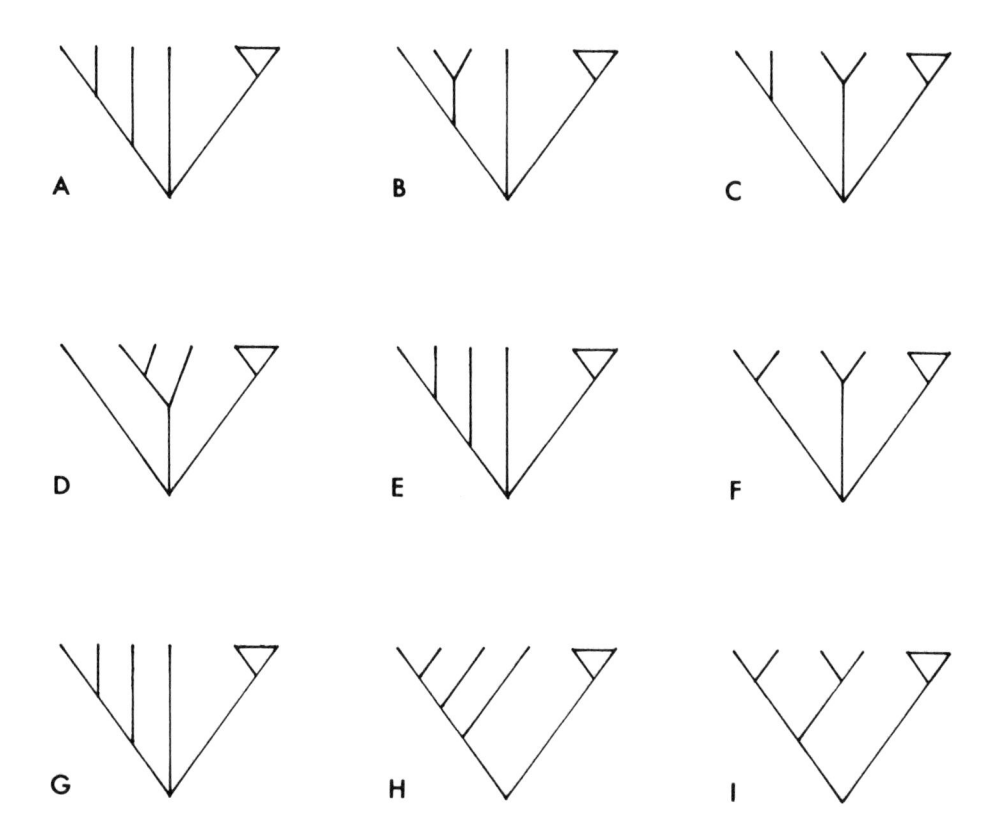

FIG. 61. Dendrograms corresponding with the nine cladograms in Figure 60 after each is rerooted at the outgroup node.

Case II (A/A/A/B): If the outgroup possessing state B attaches directly to the basal node of the cladogram (Fig. 62B), the polarity is equivocal. For all other arrangements, state A must be considered plesiomorphic.

Case III (A/A/A,B/A,B): Support for the interpretation that state B is plesiomorphic is only possible, first, if the outgroups are arranged as in Figure 62C; and second, if resolution of the relationships within the variable outgroups requires that either state B is plesiomorphic for the outgroup attaching directly to the basal node, while the other remains equivocal, or state B is plesiomorphic for both. Many arrangements of the outgroups will necessitate that state A be considered plesiomorphic, and resolution of relationships within the variable outgroups will make many arrangements equivocal. Potential determination of the plesiomorphic condition for the two variable outgroups upon resolution of relationships within these outgroups makes this case very ambiguous. In fact, there is only one arrangement in which it is impossible for the polarity inference to be equivocal (Fig. 62D).

Case IV (A/A/A,B/B): Four arrangements of four outgroups with these conditions yield equivocal evidence for polarity (Fig. 62E-H). In two of these arrangements (Fig.

TABLE 11. Summary of Polarity Inferences For Seven Cases of Character-state Distribution Among Four Outgroups of Uncertain Relationships to the Ingroup

Case	Outgroup Condition	Possible Polarity Inferences
I	A/A/A/A,B	A is plesiomorphic Polarity is equivocal*
II	A/A/A/B	A is plesiomorphic Polarity is equivocal
III	A/A/A,B/A,B	A is plesiomorphic Polarity is equivocal* B is plesiomorphic*
IV	A/A/A,B/B	A is plesiomorphic Polarity is equivocal B is plesiomorphic*
V	A/A/B/B	A is plesiomorphic Polarity is equivocal B is plesiomorphic
VI	A/A,B/A,B/A,B	A is plesiomorphic Polarity is equivocal* B is plesiomorphic*
VII	A/A,B/A,B/B	A is plesiomorphic* Polarity is equivocal B is plesiomorphic*

Note: An asterisk (*) indicates that the conclusion in question can only be reached upon resolution of relationships within one or more variable outgroups. See text for details.

62E,F), resolution of relationships within the variable outgroup may necessitate that state B be considered plesiomorphic. In all other arrangements state A must be considered plesiomorphic, although resolution of the relationships within the variable outgroup can make the situation equivocal.

Case V (A/A/B/B): In this case, only one arrangement requires that state A be considered plesiomorphic (Fig. 62I). One other arrangement requires that state B be considered plesiomorphic (Fig. 62J). For all other arrangements the polarity must be considered equivocal.

Case VI (A/A,B/A,B/A,B): Because every outgroup but one is variable, the condition

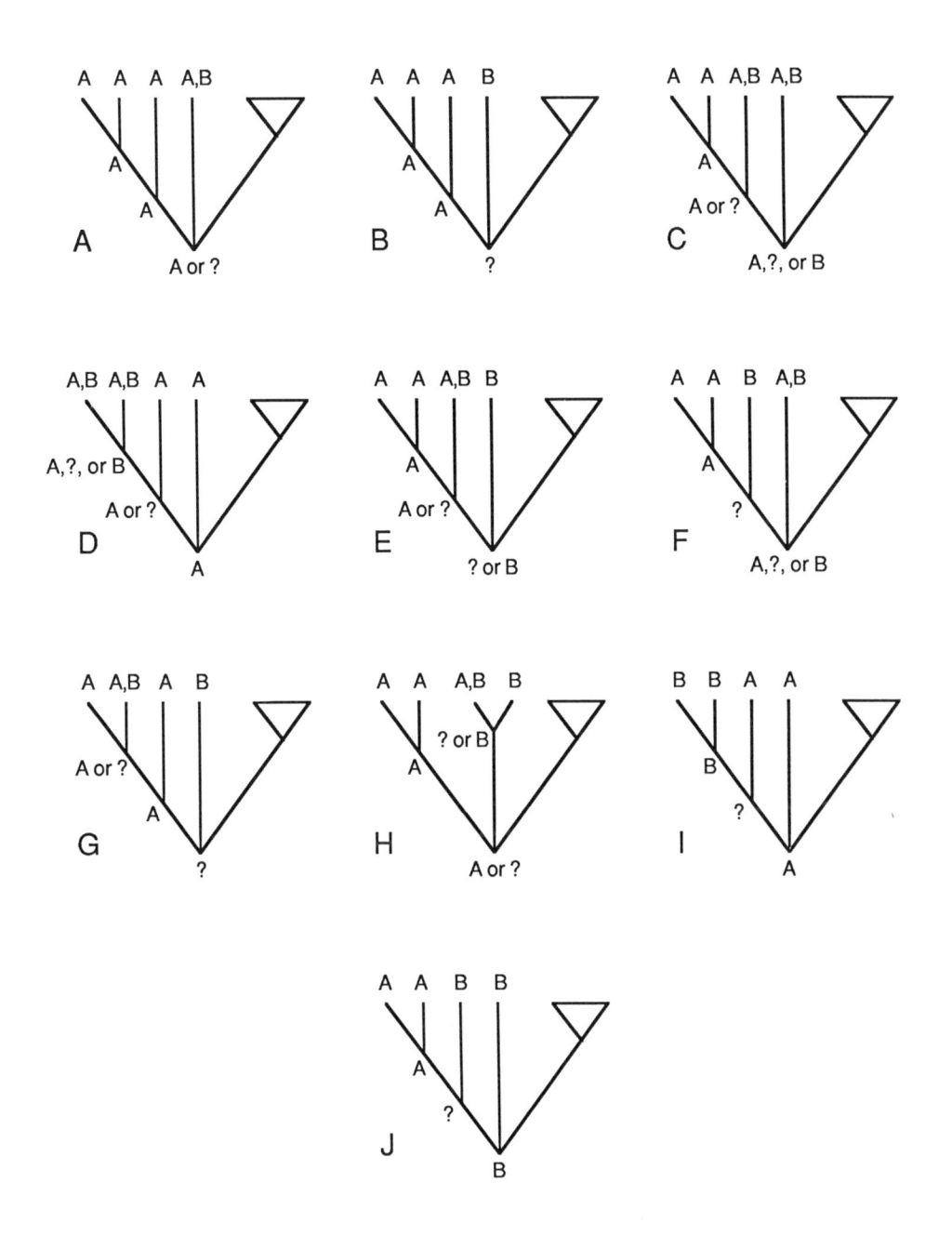

FIG. 62. Examples of polarity inferences for different arrangements of outgroup character-state distributions. See text for discussion.

of the invariable outgroup (A) must always be considered plesiomorphic. However, resolution of relationships within one or more of the variable outgroups may necessitate that either the alternative condition be considered plesiomorphic or that the polarity be considered equivocal.

Case VII (A/A,B/A,B/B): In this case, polarities are always equivocal and can only be determined by the resolution of relationships within one or more of the variable outgroups. Of course, polarities are always equivocal in the case in which all outgroups exhibit both states (A,B/A,B/A,B/A,B).

To summarize, In cases I and II (Table 11), either the more common state must be considered plesiomorphic or the situation is equivocal; the interpretation that the less common state is plesiomorphic will never be favored. In the remaining five cases there will be at least some situations in which the less common state either may or must be considered to be plesiomorphic. Therefore, I have considered the more common state to be plesiomorphic for characters with case I and II distributions, but have withheld polarity decisions on characters with case III, IV, V, VI, and VII distributions, using them only at lower hierarchical levels when certain ingroup taxa can serve as functional outgroups (Watrous and Wheeler, 1981).

Of course, not all characters fit into the cases mentioned above. For example, there are some characters with more than two states in the ingroup, and some in which one or more states found in an outgroup are not comparable to any of those seen in the ingroup. In this study, such cases are relatively rare and are discussed individually.

Appendix III
Polarity Determination for Lower Level Analysis

The polarities of 19 characters could not be determined using basiliscines, crotaphytines, morunasaurs, and oplurines as outgroups. Therefore, I attempted to determine the polarities of these characters for a less inclusive ingroup (node 3, Fig. 46), using *Brachylophus* and *Dipsosaurus* as outgroups. The problem of determining polarities for these characters is similar to that described in Appendix II, except that there are two outgroups instead of four. With only two outgroups whose relationships to the ingroup are uncertain, there are only two possible cladogram topologies (Fig. 63A,B). When rerooted at the outgroup node (Fig. 63C,D), the two resulting topologies are equivalent for the purposes of this analysis. The assessment of the condition at the outgroup node in Figure 63D must be the same as that for the node linking the two outgroups, since there are no intervening nodes. Thus, both rerooted cladograms effectively have the two outgroups attached directly to the basal (outgroup) node.

Table 12 summarizes polarity inferences for all possible arrangements of two outgroups on the rerooted cladogram (Fig. 63C) for four cases of character-state distributions among the outgroups. This exhausts the possible cases for a two-state character. Case I is unambiguous: the state found in both outgroups is plesiomorphic. In case II, state A is considered plesiomorphic, although resolution of relationships within the variable outgroup can render the polarity equivocal. In case III, the polarity is equivocal, but resolution of relationships within one or both variable outgroups may require that either state A or state B be considered plesiomorphic. Case IV is completely ambiguous: no polarity inference can be made. Because Cases I and II are the only cases in which only one of the two states can be considered plesiomorphic under all possible arrangements of the outgroups on the rerooted cladogram (Fig. 63C), I consider polarity to be determinable only for these two cases. However, none of the characters whose polarities were undeterminable at the level of all iguanines exhibits a Case II distribution. Thus, polarities are only determinable for characters with Case I distributions.

Appendix III

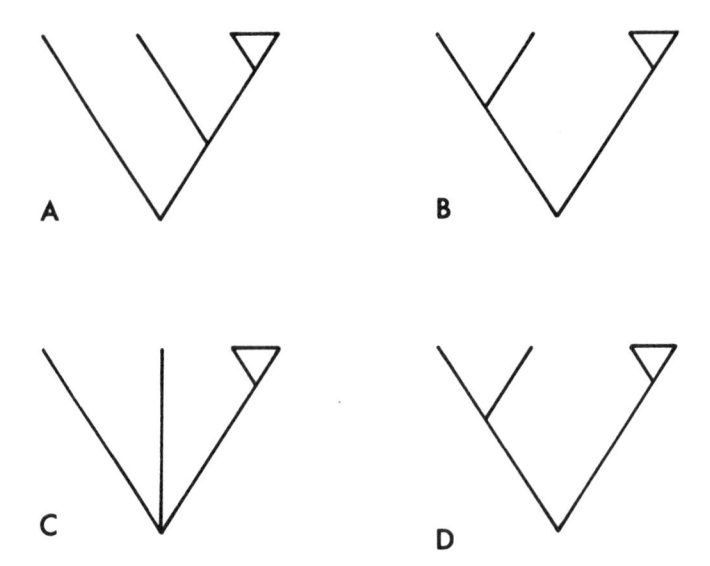

FIG. 63. All possible cladogram topologies for two unspecified outgroups and an ingroup (inverted triangle), before (A and B) and after (C and D) rerooting at the outgroup node.

TABLE 12. Summary of Polarity Inferences For Four Cases of Character-state Distribution Among Two Outgroups of Uncertain Relationships to the Ingroup

Case	Outgroup Condition	Possible Polarity Inferences
I	A/A	A is plesiomorphic
II	A/A,B	A is plesiomorphic Polarity is equivocal*
III	A,B/A,B	A is plesiomorphic* Polarity is equivocal B is plesiomorphic*
IV	A/B	Polarity is equivocal

Note: An asterisk (*) indicates that the conclusion in question can only be reached upon resolution of relationships within one or both variable outgroups.

Appendix IV
Polarity Reevaluation for Lower Level Analysis

Using *Brachylophus* and *Dipsosaurus* as outgroups to a subset of iguanines permits determination of polarities for certain characters whose polarities were undeterminable at the level of all iguanines. However, it also requires that the polarities of other characters be reassessed, since character polarities for the less inclusive group may not be identical to those for the more inclusive group. The procedure used for reassessment is similar to that used for assessing the polarities of characters whose polarities were not initially determinable (Appendix III), but differs in that a more distant outgroup must be considered (Fig. 64A,B). This more distant outgroup is actually the outgroup node from the analysis of polarities for Iguaninae as a whole. It must be considered because, unlike the case for characters whose polarities were undeterminable at the level of all iguanines, it has been assigned a character state. This additional branch also has the effect of rendering the two rerooted cladograms (Fig. 64C,D) nonequivalent.

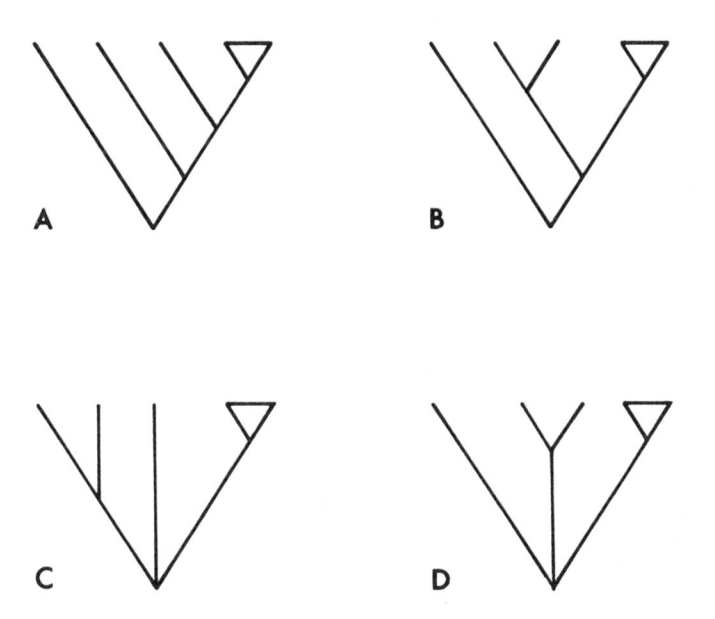

FIG. 64. All possible cladogram topologies for two unspecified near outgroups, one more remote outgroup, and an ingroup (inverted triangle), before (A and B) and after (C and D) rerooting at the outgroup node.

TABLE 13. Summary of Polarity Inferences For Six Cases of Character-state Distribution Among Two Near Outgroups Whose Precise Relationships to the Ingroup Are Unresolved, and One More Remote Outgroup Exhibiting a Fixed Character State

Case	Outgroup Condition	Possible Polarity Inferences
I	0/0	0 is plesiomorphic
II	0/0,1	0 is plesiomorphic Polarity is equivocal*
III	0,1	0 is plesiomorphic Polarity is equivocal
IV	0,1/0,1	0 is plesiomorphic Polarity is equivocal* 1 is plesiomorphic*
V	0,1/1	0 is plesiomorphic* Polarity is equivocal 1 is plesiomorphic*
VI	1/1	Polarity is equivocal 1 is plesiomorphic

Note: An asterisk (*) indicates that the conclusion can only be reached upon resolution of relationships within one or both variable outgroups. Outgroup condition refers to the near outgroups only; the remote outgroup is always assigned state zero.

Table 13 lists the possible polarity inferences for all possible arrangements of the two near outgroups on the rerooted cladograms (Fig. 64C,D) for six cases of character-state distributions among the two outgroups. This exhausts the possible character-state distributions for a two-state character. The remote outgroup is always assigned state 0, because this was inferred to be its condition based on polarity analysis at the level of all iguanines. Unlike the case for characters whose polarities were initially undeterminable (Appendix III), the numerical designations for the two outgroups are significant (e.g., 0/0 is not equivalent to 1/1), since they may or may not be identical with that of the more distant outgroup, which is always assigned state 0. Thus, there are six cases of character-state distributions rather than only four.

For characters with Case I, II, and III distributions, I have left the polarities unchanged, because evidence from the new outgroups suggests that either the character polarity for the less inclusive ingroup is identical with that for Iguaninae as a whole or that the polarity is equivocal, but in no arrangement will the reverse polarity be favored. For characters with Case IV and V distributions, I have changed the polarity assessment to undeterminable, since the new outgroup evidence is compatible with either polarity inference. For characters with Case VI distributions, I have reversed the polarity, because the new outgroups suggest that either the old polarity is incorrect for the less inclusive ingroup or the situation is equivocal.

Literature Cited

Adams, E. N.
    1972. Consensus techniques and the comparison of taxonomic trees. Syst. Zool. 21(4):390-397.

Avery, D. F., and W. W. Tanner
    1964. The osteology and myology of the head and thorax regions of the *obesus* group of the genus *Sauromalus* Dumeril (Iguanidae). Brigham Young Univ. Sci. Bull., Biol. ser. 5(3):1-30.
    1971. Evolution of the iguanine lizards (Sauria: Iguanidae) as determined by osteological and myological characters. Brigham Young Univ. Sci. Bull., Biol. ser. 12(3):1-79.

Axtell, R. W.
    1958. A monographic revision of the iguanid genus *Holbrookia*. Ph.D. dissertation, Univ. Texas, Austin, 222 pp.

Bailey, J. W.
    1928. A revision of the lizards of the genus *Ctenosaura*. Proc. U.S. Natl. Mus. 73(12):1-58.

Baird, S. F., and C. Girard
    1852. Characteristics of some new reptiles in the museum of the Smithsonian Institution. Proc. Acad. Nat. Sci. Philadelphia 6:125-129.

Barbour, T.
    1919. A new rock iguana from Porto Rico. Proc. Biol. Soc. Wash. 32:145-148.

Barbour, T., and G. K. Noble
    1916. A revision of the lizards of the genus *Cyclura*. Bull. Mus. Comp. Zool., Harvard, 60(4):139-164 + 15 plates.

Bell, T.
    1825. On a new genus of Iguanidae. Zool. J. 2:204-207 + 1 plate.

Bellairs, A. d'A., and S. V. Bryant
    1985. Autotomy and regeneration in reptiles. Pp. 301-410 in Biology of the Reptilia, Vol. 15, Development B (C. Gans and F. Billett, eds.). John Wiley, New York.

Boulenger, G. A.
    1884. Synopsis of the families of existing Lacertilia. Ann. Mag. Nat. Hist., ser. 5, 24:117-122.
    1885. Catalogue of the lizards in the British Museum (Natural History), Vol. 2. British Museum (Nat. Hist.), London.

1890. On the distinctive cranial characters of the iguanoid lizards allied to *Iguana*. Ann. Mag. Nat. Hist., ser. 6, 6(35):412-414.

Brongniart, A.
   1800. Essai d'une classification naturelle des reptiles. Bull. Soc. Philom., Paris, 2(36):89-91 + 1 plate.
   1805. Essai d'une classification naturelle des reptiles. Baudouin, Imprimeur de l'Institut National, Paris.

Burghardt, G. M., and A. S. Rand (eds.)
   1982. Iguanas of the world: Their behavior, ecology, and conservation. Noyes Publ., Park Ridge, N. J., xix + 472 pp.

Camp, C. L.
   1923. Classification of the lizards. Bull. Amer. Mus. Nat. Hist. 48(11):289-481.

Carlquist, S.
   1974. Island biology. Columbia Univ. Press, New York, 660 pp.

Carpenter, C. C.
   1966. The marine iguana of the Galápagos Islands, its behavior and ecology. Proc. Calif. Acad. Sci. 34(6):329-376.

Carroll, R. L.
   1977. The origin of lizards. Pp. 359-396 *in* Problems in vertebrate evolution (S. M. Andrews, R. S. Miles, and A. D. Walker, eds.). Linn. Soc. Symp., ser. 4.

Cliff, F. S.
   1958. A new species of *Sauromalus* from Mexico. Copeia 1958(4):259-261.

Cohen, M. M., C. C. Huang, and H. F. Clark
   1967. The somatic chromosomes of 3 lizard species: *Gekko gekko*, *Iguana iguana*, and *Crotaphytus collaris*. Experientia 23:769-771.

Congdon, J. D., L. J. Vitt, and W. W. King
   1974. Geckos: Adaptive significance and energetics of tail autotomy. Science 184:1379-1380.

Cope, E. D.
   1886. On the species of Iguaninae. Proc. Amer. Philos. Soc. 23(122):261-271.
   1900. The crocodilians, lizards, and snakes of North America. Ann. Rept. U.S. Natl. Mus., 1898, part 2:151-1270 + 36 plates.

Cuvier, G.
   1817. Le règne animal distribué d'après son organisation, pour servir de base a l'histoire naturelle des animaux et d'introduction a l'anatomie comparée, Tome 2. Chez Déterville, Libraire, Paris.
   1829. Le règne animal distribué d'après son organisation, pour servir de base a l'histoire naturelle des animaux et d'introduction a l'anatomie comparée. Tome 2. Chez Déterville, Libraire, Paris.

1831. The animal kingdom arranged in conformity with its organization, with additional descriptions by Edward Griffith and others, Vol. 9, The Class Reptilia. Whittaker, Treacher, and Co., London.

Darwin, C.
1835. The journal of a voyage in H.M.S. Beagle. Genesis Publ., Surrey, England, 1979.
1859. On the origin of species by means of natural selection. John Murray, London.

Daudin, F. M.
1805. Histoire naturelle, generale et particulière des reptiles. Imprimerie de F. Dufart, Paris.

Dawson, W. R., G. A. Bartholomew, and A. F. Bennett
1977. A reappraisal of the aquatic specializations of the Galapagos marine iguana (*Amblyrhynchus cristatus*). Evolution 31(4):891-897.

Dee Boersma, P.
1983. An ecological study of the Galapagos marine iguana. Pp. 157-176 in Patterns of evolution in Galapagos organisms (R. I. Bowman, M. Benson, and A. E. Leviton, eds.). Pacific Div. Amer. Assoc. Adv. Sci.

de Queiroz, K.
1982. The scleral ossicles of sceloporine iguanids: A reexamination with comments on their phylogenetic significance. Herpetologica 38(2):302-311.
1985. The ontogenetic method for determining character polarity and its relevance to phylogenetic systematics. Syst. Zool. 34(3):280-299.
1987. Systematics and the Darwinian Revolution. Phil. Sci., in press.

Donoghue, M. J., and P. Cantino
1984. The logic and limitations of the outgroup substitution approach to cladistic analysis. Syst. Bot. 9(2):192-202.

Duellman, W. E.
1965. Amphibians and reptiles from the Yucatan Peninsula, México. Univ. Kansas Publ. Mus. Nat. Hist. 15(12):577-614.

Duellman, W. E., and A. S. Duellman
1959. Variation, distribution, and ecology of the iguanid lizard *Enyaliosaurus clarki* of Michoacan, Mexico. Occ. Pap. Mus. Zool. Univ. Michigan 598:1-11.

Duméril, A.
1856. Description des reptiles nouveaux ou imparfaitement connus de la collection du Muséum d'Histoire Naturelle et remarques sur la classification et les caractères des reptiles. Arch. Mus. Hist. Nat., Paris 8:56-588.

Duméril, A. M. C., and G. Bibron
1837. Erpétologie générale: Ou histoire naturelle complète des reptiles. Librairie Encyclopédique de Roret, Paris.

Dunn, E. R.
    1934. Notes on *Iguana*. Copeia 1934(1):1-4.

Dunson, W. A.
    1969. Electrolyte excretion by the salt gland of the Galápagos marine iguana. Amer. J. Physiol. 216(4):995-1002.
    1976. Salt glands in reptiles. Pp. 413-445 *in* Biology of the Reptilia, Vol. 5, Physiology (C. Gans and W. R. Dawson, eds.). Academic Press, New York.

Earle, A. M.
    1962. The middle ear of the genus *Uma* compared to those of other sand lizards. Copeia 1962(1):185-188.

Edmund, A. G.
    1960. Tooth replacement phenomena in the lower vertebrates. Royal Ont. Mus., Life Sci. Div. Contr. 52:1-190.

Eibl-Eibesfeldt, I.
    1961. Galapagos: The Noah's Ark of the Pacific. Doubleday, New York, 192 pp.

Eldredge, N., and J. Cracraft
    1980. Phylogenetic patterns and the evolutionary process. Columbia Univ. Press, New York, 349 pp.

Estes, R.
    1963. Early Miocene salamanders and lizards from Florida. Quart. J. Florida Acad. Sci. 26(3):234-256.
    1983. Sauria terrestria, Amphisbaenia. Handbuch der Palaoherpetologie, Teil 10A. Gustav Fischer, Stuttgart, xii + 249 pp.

Estes, R., K. de Queiroz, and J. A. Gauthier
    1988. Phylogenetic relationships within squamate reptiles. *In* Proceedings of a symposium on the phylogenetic relationships of the lizard families (R. Estes and G. Pregill, eds.), Stanford Univ. Press, in press.

Estes, R., and L. I. Price
    1973. Iguanid lizard from the Upper Cretaceous of Brazil. Science 180:748-751.

Etheridge, R.
    1959. The relationships of the anoles (Reptilia: Sauria: Iguanidae) an interpretation based on skeletal morphology. Ph.D. dissertation, Univ. Michigan, 236 pp.
    1962. Skeletal variation in the iguanid lizard *Sator grandaevus*. Copeia 1962(3):613-619.
    1964a. The skeletal morphology and systematic relationships of sceloporine lizards. Copeia 1964(4):610-631.
    1964b. Late Pleistocene lizards from Barbuda, British West Indies. Bull. Florida State Mus., Biol. Sci. 9:43-75.

1965a. Fossil lizards from the Dominican Republic. Quart. J. Florida Acad. Sci. 28(1):83-105.
1965b. The abdominal skeleton of lizards in the family Iguanidae. Herpetologica 21(3):161-168.
1965c. Pleistocene lizards from New Providence. Quart. J. Florida Acad. Sci. 28(1):349-358.
1966. The systematic relationships of West Indian and South American lizards referred to the iguanid genus *Leiocephalus*. Copeia 1966(1):79-91.
1967. Lizard caudal vertebrae. Copeia 1967(4):699-721.
1982. Checklist of iguanine and Malagasy iguanid lizards. Pp. 7-37 *in* Iguanas of the world: Their behavior, ecology, and conservation (G. M. Burghardt and A. S. Rand, eds.). Noyes Publ., Park Ridge, N. J.

Etheridge, R., and K. de Queiroz
1988. A phylogeny of Iguanidae. *In* Proceedings of a symposium on the phylogenetic relationships of the lizard families (R. Estes and G. Pregill, eds.), Stanford Univ. Press, in press.

Farris, J. S.
1970. Methods for computing Wagner trees. Syst. Zool. 19(1):83-92.
1982. Outgroups and parsimony. Syst. Zool. 31(3):328-334.

Fitch, H. S., and J. Hackforth-Jones
1983. *Ctenosaura similis*. Pp. 394-396 *in* Costa Rican natural history (D. H. Janzen, ed.). Univ. Chicago Press.

Fitzinger, L. I.
1826. Nue Classification der Reptilien. Verlage von J. G. Heubner, Vienna.
1843. Systema Reptilium. Braumüller et Seidel Bibliopolas, Vindobonae. Reprinted by Soc. Study Amphib. Rept., 1973.

Fries, C., C. W. Hibbard, and D. H. Dunkle
1955. Early Cenozoic vertebrates in the red conglomerate at Guanajuato, Mexico. Smith. Misc. Coll. 123(7):1-25.

Gates, G. O.
1968. Geographical distribution and character-analysis of the iguanid lizard *Sauromalus obesus* in Baja California, Mexico. Herpetologica 24(4):285-288.

Gauthier, J. A., R. Estes, and K. de Queiroz
1988. A phylogenetic analysis of Lepidosauromorpha. *In* Proceedings of a symposium on the phylogenetic relationships of the lizard families (R. Estes and G. Pregill, eds.), Stanford Univ. Press, in press.

Gibbons, J. R. H.
1981. The biogeography of *Brachylophus* (Iguanidae) including the description of a new species, *B. vitiensis*, from Fiji. J. Herpetol. 15(3):255-273.

Gicca, D.
1983. *Enyaliosaurus quinquecarinatus*. Cat. Amer. Amphib. Rept. 329:1-2.

Gilmore, C. W.
 1928. Fossil lizards of North America. Mem. Natl. Acad. Sci. 22:1-197.

Goin, C. J., O. B. Goin, and G. R. Zug
 1978. Introduction to herpetology. Third ed. W. H. Freeman, San Francisco, 378 pp.

Gorman, G. C.
 1973. The chromosomes of the Reptilia, a cytotaxonomic interpretation. Pp. 349-424 *in* Cytotaxonomy and vertebrate evolution (A. B. Chiarelli and E. Capanna, eds.). Academic Press, New York.

Gorman, G. C., L. Atkins, and T. Holzinger
 1967. New karyotypic data on 15 genera of lizards in the Family Iguanidae, with a discussion of taxonomic and cytological implications. Cytogenetics 6:286-299.

Gorman, G. C., A. C. Wilson, and M. Nakanishi
 1971. A biochemical approach to the study of reptilian phylogeny: Evolution of serum albumin and lactic dehydrogenase. Syst. Zool. 20(2):167-185.

Gray, J. E.
 1831a. A synopsis of the species of the class Reptilia. Pp. 1-110, paginated separately, *in* G. Cuvier, The animal kingdom arranged in conformity with its organization, with additional descriptions by Edward Griffith and others, Vol. 9, The Class Reptilia. Whittaker, Treacher, and Co., London.
 1831b. Description of a new species of *Amblyrynchus* of Mr. Bell, in the British Museum. Zool. Misc. 10:6. Reprinted by Soc. Study Amphib. Rept., 1971.
 1845. Catalogue of the specimens of lizards in the collection of the British Museum. Edward Newman, London, xxviii + 289 pp.

Gugg, W.
 1939. Der Skleralring der plagiotremen Reptilien. Zool. Jb. (Anat.) 65:339-416.

Hallowell, E.
 1854. Descriptions of new reptiles from California. Proc. Acad. Nat. Sci. Philadelphia 7:91-97.

Hamasaki, D. I.
 1968. Properties of the parietal eye of the green iguana. Vision Res. 8:591-599.
 1969. Spectral sensitivity of the parietal eye of the green iguana. Vision Res. 9:515-523.

Hardy, L. M., and R. W. McDiarmid
 1969. The amphibians and reptiles of Sinaloa, México. Univ. Kansas Publ. Mus. Nat. Hist. 18(3):39-252.

Harlan, R.
 1824. Description of two species of Linnaean *Lacerta*, not before described, and construction of the new genus *Cyclura*. J. Acad. Nat. Sci. Philadelphia 4:242-251 + 2 plates.

Heller, E.
  1903. Papers from the Hopkins Stanford Galapagos expedition, 1898-1899, XIV. Reptiles. Proc. Wash. Acad. Sci. 5:39-98.

Hennig, W.
  1966. Phylogenetic systematics. Univ. Illinois Press, Urbana, 263 pp.

Hidalgo, H.
  1980. *Enyaliosaurus quinquecarinatus* (Gray) and *Leptodeira nigrofasciata* Günther in El Salvador. Herpetol. Rev. 11(2):41-43.

Higgins, P. J.
  1977. Immunodiffusion comparisons of the serum albumins of marine and land iguanas from different islands in the Galapagos Archipelago. Canadian J. Zool. 55:1389-1392.
  1978. The Galápagos iguanas: Models of reptilian differentiation. Bioscience 28(8):512-515.

Higgins, P. J., and C. S. Rand
  1974. A comparative immunochemical study of the serum proteins of several Galapagos iguanids. Comp. Biochem. Physiol. 49A:347-355.
  1975. Comparative immunology of Galapagos iguana hemoglobins. J. Exp. Zool. 193:391-397.

Hoffstetter, R.
  1946. Faune du gisement précolombien d'Anse-Belleville (Martinique). Mem. Mus. Nation. Hist. Nat., Paris 22:1-18.
  1970. Vertebrados cenozoicos de Ecuador. Actas IV Congr. Latinoamer. de Zool., Carácas, 2:955-970.

Hoffstetter, R., and J.-P. Gasc
  1969. Vertebrae and ribs of modern reptiles. Pp. 201-310 *in* Biology of the Reptilia, Vol. 1, Morphology A (C. Gans, A. d'A. Bellairs, and T. S. Parsons, eds.). Academic Press, New York.

Hotton, N.
  1955. A survey of adaptive relationships of dentition to diet in North American Iguanidae. Amer. Midl. Nat. 53(1):88-114.

Iverson, J. B.
  1980. Colic modifications in iguanine lizards. J. Morph. 163:79-93.
  1982. Adaptations to herbivory in iguanine lizards. Pp. 60-76 *in* Iguanas of the world: Their behavior, ecology, and conservation (G. M. Burghardt and A. S. Rand, eds.). Noyes Publ., Park Ridge, N. J.

Jenkins, F. A., and G. E. Goslow
  1983. The functional anatomy of the shoulder of the savannah monitor lizard (*Varanus exanthematicus*). J. Morph. 175:195-216.

Kluge, A. G., and J. S. Farris
  1969. Quantitative phyletics and the evolution of anurans. Syst. Zool. 18(1):1-32.

Lacépède, B. G.
    1788.   Histoire naturelle des quadrupèdes ovipares et des serpens. 2 Vols. Hotel de Thou, Paris.

Langebartel, D.
    1953.   The reptiles and amphibians. Pp. 91-108 *in* Faunal and archaeological researches in Yucatan caves (R. T. Hatt, ed.). Cranbrook Inst. Sci. Bull. 33.

Latreille, M.
    1825.   Familles naturelles du règne animal, exposées succinctement et dans un ordre analytique, avec l'indication de leurs genres. J.-B. Baillière, Libraire, Paris.

Laurenti, J. N.
    1768.   Specimen medicum, exhibens synopsin reptilium emendatam cum experimentis circa venena et antidota reptilium austriacorum. Joan. Thom. Nob. de Trattnern, Vienna.

Lazell, J. D.
    1973.   The lizard genus *Iguana* in the Lesser Antilles. Bull. Mus. Comp. Zool., Harvard, 145(1):1-28.

Lécuru, S.
    1968a.   Remarques sur le scapulo-coracoïde des lacertiliens. Ann. Sci. Nat. Zool., Paris, 10:475-510.
    1968b.   Étude des variations morphologiques du sternum, des clavicules et de l'interclavicule des lacertiliens. Ann. Sci. Nat. Zool., Paris, 10:511-544.

Linnaeus, C.
    1758.   Systema naturae, per regna tria naturae, secundum classes, ordines, genera, species cum characteribus, differentiis, synonymis, locis. 10th ed. Laurentii Salvii, Holmiae, 824 pp.

Maddison, W. P., M. J. Donoghue, and D. R. Maddison
    1984.   Outgroup analysis and parsimony. Syst. Zool. 33(1):83-103.

Merrem, B.
    1820.   Versuch eines Systems der Amphibien. Johann Christian Krieger, Marburg.

Meyer, J. R., and L. D. Wilson
    1973.   A distributional checklist of the turtles, crocodilians, and lizards of Honduras. Contr. Sci. Nat. Hist. Mus. Los Angeles Co. 244:1-39.

Miller, G. S.
    1918.   Mammals and reptiles collected by Theodoor de Booy in the Virgin Islands. Proc. U.S. Natl. Mus. 54:507-511.

Mittleman, M. B.
    1942.   A summary of the iguanid genus *Urosaurus*. Bull. Mus. Comp. Zool., Harvard, 91(2):105-181 + 16 plates.

Montanucci, R. R.
    1968.   Comparative dentition in four iguanid lizards. Herpetologica 24(4):305-315.

Moody, S. M.
    1980.   Phylogenetic and historical biogeographical relationships of the genera in the family Agamidae (Reptilia: Lacertilia). Ph.D. dissertation, Univ. Michigan, xv + 373 pp.
    1982.   Cladistic relationships within the Iguania. Paper presented at a symposium on the phylogenetic relationships of the lizard families, Annual meeting of the American Society of Zoologists, Louisville, Ky.

Nelson, G., and N. I. Platnick
    1981.   Systematics and biogeography: Cladistics and vicariance. Columbia Univ. Press, New York, 567 pp.

Norell, M. A.
    1983.   Late Neogene lizards from the Anza Borrego Desert, San Diego County, California. M.S. thesis, San Diego State Univ., 199 pp.
    1986.   Late Pleistocene lizards from Kokoweef Cave, San Bernardino County, California. Copeia 1986(1):244-246.

Oelrich, T. M.
    1956.   The anatomy of the head of *Ctenosaura pectinata* (Iguanidae). Misc. Publ. Mus. Zool. Univ. Michigan 94:1-122 + 59 figs.

Olson, E. C.
    1937.   A Miocene lizard from Nebraska. Herpetologica 1(4):111-112.

Patterson, C.
    1982.   Morphological characters and homology. Pp. 21-74 *in* Problems of phylogenetic reconstruction (K. A. Joysey and A. E. Friday, eds.), Syst. assoc. spec. vol. no. 21. Academic Press, New York.

Paull, D., E. E. Williams, and W. P. Hall
    1976.   Lizard karyotypes from the Galapagos Islands: Chromosomes in phylogeny and evolution. Breviora Mus. Comp. Zool., Harvard, 441:1-31.

Peters, J. A., and R. Donoso-Barros
    1970.   Catalogue of the Neotropical Squamata, part 2, Lizards and amphisbaenians. U.S. Natl. Mus. Bull. 297:viii + 293 pp.

Peterson, J. A.
    1973.   Adaptation for arboreal locomotion in the shoulder region of lizards. Ph.D. dissertation, Univ. Chicago, 584 pp.
    1984.   The microstructure of the scale surface in iguanid lizards. J. Herpetol. 18(4):437-467.

Pregill, G.
    1981.   Late Pleistocene herpetofaunas from Puerto Rico. Univ. Kansas Mus. Nat. Hist. Misc. Publ. 71:1-72.

1982. Fossil amphibians and reptiles from New Providence Island, Bahamas. Pp. 8-21 *in* Fossil vertebrates from the Bahamas (S. L. Olson, ed.). Smith. Contr. Paleobiol. 48.

Ray, C. E.
 1964. A small assemblage of vertebrate fossils from Spring Bay, Barbados. J. Barbados Mus. Hist. Soc. 31(1):11-22.
 1965. Variation in the number of marginal tooth positions in three species of iguanid lizards. Breviora Mus. Comp. Zool., Harvard, 236:1-15.

Ray, C. E., and E. E. Williams
 Remarks on the taxonomy of the iguanid genus *Ctenosaura* and a key to its species. Unpublished.

Renous, S.
 1979. Application des principes cladistiques à la phylogénèse et la biogéographie des lacertiliens. Gegenbaurs Morph. Jahrb., Leipzig, 125:376-432.

Renous-Lécuru, S.
 1973. Morphologie comparée du carpe chez les lepidosauriens actuels (rhynchocéphales, lacertiliens, amphisbéniens). Gegenbaurs Morph. Jahrb., Leipzig, 119:727-766.

Rieppel, O.
 1978. Streptostyly and muscle function in lizards. Experientia 34:776-777.

Robinson, M. D.
 1972. Chromosomes, protein polymorphism and systematics of insular chuckwalla lizards (genus *Sauromalus*) in the Gulf of California, Mexico. Ph.D. dissertation, Univ. Arizona, 71 pp.
 1974. Chromosomes of the insular species of chuckwalla lizards (genus *Sauromalus*) in the Gulf of California, Mexico. Herpetologica 30(2):162-167.

Savage, J. M.
 1958. The iguanid lizard genera *Urosaurus* and *Uta*, with remarks on related groups. Zoologica 43(2):41-54.

Schmidt-Nielsen, K., and R. Fange
 1958. Salt glands in marine reptiles. Nature 182:783-785.

Schwartz, A., and M. Carey
 1977. Systematics and evolution in the West Indian iguanid genus *Cyclura*. Stud. Fauna of Curacao and Other Caribbean Islands 53:15-97.

Shaw, C. E.
 1945. The chuckwallas, genus *Sauromalus*. Trans. San Diego Soc. Nat. Hist. 10(15):269-306.

Shaw, G.
 1802. General zoology, vol. 3, part. 1, Amphibia. G. Kearsley, London, 312 pp.

Smith, H. M.
- 1946. Handbook of lizards: Lizards of the United States and Canada. Cornell Univ. Press, Ithaca, N. Y., 557 pp.
- 1949. Miscellaneous notes on Mexican lizards. J. Wash. Acad. Sci. 39(1):34-43.
- 1972. The Sonoran subspecies of the lizard *Ctenosaura hemilopha*. Gt. Basin Nat. 32(2):104-111.

Smith, H. M., and E. H. Taylor
- 1950. An annotated checklist and key to the reptiles of Mexico exclusive of the snakes. Bull. U.S. Natl. Mus. 199:1-253.

Smith, K. K.
- 1980. Mechanical significance of streptostyly in lizards. Nature 283:778-779.

Soulé, M., and A. J. Sloan
- 1966. Biogeography and distribution of the reptiles and amphibians on islands in the Gulf of California, Mexico. Trans. San Diego Soc. Nat. Hist. 14(11):137-156.

Steadman, D.
- 1981. Vertebrate fossils in lava tubes in the Galápagos Islands. Proc. Eighth Int. Congr. Speleology 549-550.

Steadman, D. W., D. R. Watters, E. J. Reitz, and G. K. Pregill
- 1984. Vertebrates from archaeological sites on Montserrat, West Indies. Ann. Carnegie Mus. 53(1):1-29.

Stebbins, R. C.
- 1948. Nasal structure in lizards with reference to olfaction and conditioning of the inspired air. Amer. J. Anat. 83(2):183-222.
- 1966. A field guide to western reptiles and amphibians. Houghton Mifflin, Boston, 279 pp.

Stejneger, L.
- 1901. On a new species of spiny-tailed iguana from Utilla Island, Honduras. Proc. U.S. Natl. Mus. 23(1217):467-468.

Stevens, M. S.
- 1977. Further study of Castolon local fauna (early Miocene), Big Bend National Park, Texas. Pearce-Sellards ser., Texas Memorial Mus., 28:1-69.

Swinton, W. E.
- 1937. *Iguana* remains from Barbados. Ann. Mag. Nat. Hist. 19:306-307.

Tanner, W. W., and D. F. Avery
- 1964. A new *Sauromalus obesus* from the Upper Colorado Basin of Utah. Herpetologica 20(1):38-42.

Thomas, R.
- 1966. A reassessmant of the herpetofauna of Navassa Island. J. Ohio Herpetol. Soc. 5(3):73-89.

Thornton, I.
  1971.   Darwin's islands: A natural history of the Galápagos. Natural History Press, New York, 322 pp.
Tracy, R. C., and K. A. Christian
  1985.   Are marine iguana tails flattened? Brit. J. Herpetol. 6:434-435.
Troyer, K.
  1983.   The biology of iguanine lizards: Present status and future directions. Herpetologica 39(3):317-328.
Turner, F. B., P. A. Medica, R. I. Jennrich, and B. G. Maza
  1982.   Frequencies of broken tails among *Uta stansburiana* in southern Nevada and a test of the predation hypothesis. Copeia 1982(4):835-840.
Underwood, G.
  1970.   The eye. Pp. 1-97 *in* Biology of the Reptilia, Vol. 2, Morphology B (C. Gans and T. S. Parsons, eds.). Academic Press, New York.
Van Denburgh, J.
  1922.   The reptiles of western North America, Vol. 1, Lizards. Occ. Pap. Calif. Acad. Sci. 10:1-611.
Wagler, J.
  1830.   Natürliches System der Amphibien, mit vorangehender Classification der Säugthiere und Vögel. J. G. Cotta'schen Buchhandlung, München, Stuttgart, and Tübingen.
Watrous, L. E., and Q. D. Wheeler
  1981.   The outgroup comparison method of character analysis. Syst. Zool. 30(1):1-11.
Wiegmann, A. F. A.
  1828.   Beiträge zur Amphibienkunde. Isis von Oken 21:364-383.
  1834.   Herpetologia mexicana seu descriptio amphibiorum Novae Hispaniae. C. G. Lunderitz, Berlin. Reprinted by Soc. Study Amphib. Rept., 1969.
Wiley, E. O.
  1979.   Ancestors, species, and cladograms--remarks on the symposium. Pp. 211-225 *in* Phylogenetic analysis and paleontology (J. Cracraft and N. Eldredge, eds.). Columbia Univ. Press, New York.
  1981.   Phylogenetics: The theory and practice of phylogenetic systematics. John Wiley, New York, 439 pp.
Wilson, L. D., and D. E. Hahn
  1973.   The herpetofauna of the Islas de la Bahía, Honduras. Bull. Florida State Mus., Biol. Sci. 17(2):93-150.
Wing, E. S., C. A. Hoffman, Jr., and C. E. Ray
  1968.   Vertebrate remains from Indian sites on Antigua, West Indies. Caribbean J. Sci. 8(3-4):123-139.
Woodley, J. D.
  1980.   Survival of the Jamaican iguana, *Cyclura collei*. J. Herpetol. 14(1):45-49.

Wyles, J. S., and V. M. Sarich
- 1983. Are the Galapagos iguanas older than the Galapagos? Pp. 177-186 *in* Patterns of evolution in Galapagos organisms (R. I. Bowman, M. Benson, and A. E. Leviton, eds.). Pacific Div. Amer. Assoc. Adv. Sci.

Zug, G. R.
- 1971. The distribution and patterns of the major arteries of the iguanids and comments on the intergeneric relationships of iguanids (Reptilia: Lacertilia). Smith. Contr. Zool. 83:1-23.

Forsyth Library

Other Volumes Available
University of California Publications in Zoology

Vol. 112. Ned K. Johnson. *Character Variation of Evolution of Sibling Species in the Empidonax Difficilis-Flavescens Complex (Aves: Tyrannidae).* ISBN 0-520-09799-5.

Vol. 113. Thomas R. Howell. *Breeding Biology of the Egyptian Plover Pluvianus Argyptius (Aves: Glareolidae).* ISBN 0-520-03804-5.

Vol. 114. Blair Csuti. *Type Specimens of Recent Mammals in the Museum of Vertebrate Zoology, University of California, Berkeley.* ISBN 0-520-09622-3.

Vol. 115. Peter B. Moyle et al. *Distribution and Ecology of Stream Fishes of the Sacramento-San Joaquin Drainage System, California.* ISBN 0-520-09650-9.

Vol. 116. Russell Greenberg. *The Winter Exploitation Systems of Bay-breasted and Chestnut-sided Warblers in Panama.* ISBN 0-520-09670-3.

Vol. 117. Marina Cords. *Mixed-Species Association of Cercopithecus Monkeys in the Kakamega Forest, Kenya.* ISBN 0-520-09717-3.

ISBN 0-520-09730-0